The Open University

Mathematics/Science/Technology:
An Inter-Faculty Second Level Course

Elementary Mathematics for Science and Technology

Unit 1   SETS, MAPPINGS AND
SEQUENCES

Unit 2   FUNCTIONS AND LIMITS

Prepared by the Course Team

The Open University Press

The Open University Press Walton Hall Bletchley Bucks

First published 1972

Copyright © 1972 The Open University

Designed by the Media Development Group of the Open University.

Printed in Great Britain by
Martin Cadbury Printing Group

SBN 335 012108

This text forms part of the correspondence element of an Open University
Second Level Course. The complete list of units in the course is given
at the end of this text.

For general availability of supporting material referred to in this text,
please write to the Director of Marketing, The Open University, Walton
Hall, Bletchley, Buckinghamshire.

Further information on Open University courses may be obtained from
the Admissions Office, The Open University, P.O. Box 48, Bletchley,
Buckinghamshire.

1.1

Unit 1 Sets, Mappings and Sequences

Contents

Bibliography

This bibliography is intended to furnish a list of books which you may find useful after you have completed the study of the core material in a particular unit. They should not be regarded as alternatives to the set book. Each will provide a useful source of additional examples and exercises. When using one of these books you should take care in interpreting the material, as the notation and terminology may differ from ours.

Books Which Cover a Large Part of the Course Material

A. Jeffrey, *Mathematics for Engineers and Scientists*, Applications for Mathematics Series (Nelson, 1969).

This book presents a broad and modern introduction to mathematics for students of engineering and the physical sciences. The book pays particular attention to offering reasonable proofs of the main mathematical results as well as to the application of such results.

The text places on a firm formal foundation many of the results which are discussed intuitively in our course. If you have mastered our intuitive approach to a particular concept, you may like to follow this up by studying the formal basis for the concept concerned. You should certainly find the numerous examples and exercises (many with outline solutions) helpful when you are revising the course material.

The material covered in the book is rather broader than in our course; in particular, most of the material in Chapters 10, 11 and 15 is beyond the scope of our present course. Many of the other chapters take the treatment of a particular topic to greater depth than we do. If the book is used for revision and further study, then you can use your own discretion as to the depth to which you follow a particular argument. The book is well cross-referenced so that it is easy to discover the interdependence of various concepts.

M. Bruckheimer, N. W. Gowar and R. E. Scraton, *Mathematics for Technology: A New Approach* (Chatto and Windus, 1968).

This book presents an introduction to mathematics similar in approach to that which we have adopted. For this reason you should find the book of particular value when you are reviewing the ideas which we discuss in the course. The material is developed by means of structured sets of examples and exercises (many with solutions) supported by explanatory notes on the concepts involved. This approach contributes to the book's value as a revision text.

Little of the material in this book is beyond the scope of our present course. However, the concepts are developed in a different order from ours. Consequently, if you make reference to this text early on during your work, you must take care to avoid selecting material for study which depends upon concepts which you have not yet encountered. In most cases you can pass over unfamiliar concepts without affecting the value which can be derived from working a particular group of examples and exercises.

Books Which Cover Part of the Course Material

There are, of course, a large number of books which cover the concepts which we treat in our course material. We have selected just one text on Calculus and one text on Algebra in which the approach and depth of treatment is similar to that we have adopted for our course.

K. De Leeuw, *Calculus*, Harbrace College Mathematics Series (Harcourt, Brace and World, 1966).

This book presents a comprehensive introduction to the concepts and techniques of the differential and integral calculus of functions of one variable. The book presents a careful statement of each of the basic definitions and theorems; from then on much of the development is informal and intuitive as in our own course. The book contains a chapter on the foundations of the calculus in which a number of arguments are formally justified. Where the proof of a particular result is beyond the scope of an introductory course, reference is given to books in which the proof may be found.

All the basic definitions and theorems are supported by explanatory notes and examples, and a comprehensive guide to further reading is provided. The book contains a very wide selection of exercises (many with outline solutions) which have been designed to develop your intuition and to lead you to discover facts about the calculus for yourself. With the exception of some of the material on the foundations of the calculus, all the material covered in the book falls within the scope of our course. Since both the coverage and level of presentation are similar to our own, this makes the book particularly valuable as a revision text.

Hans Liebeck, *Introductory Algebra for Scientists and Engineers* (John Wiley, 1969).

This book presents an introduction to those algebraic concepts which are basic to the mathematical analysis of problems from science and technology. Particular emphasis is laid upon the applications of modern algebra. The level of the presentation is similar to that in our own course, although less emphasis is placed on the mathematical structure approach to abstract algebra. The material covered is broader than that covered in our course. In particular, some of the material in Chapter 5 and all the material in Chapters 13–18 and Part II are outside the scope of our course. This material provides useful further reading in the study of algebra. The book does not cover the concept of complex numbers (which we have treated algebraically); however, adequate coverage of this topic can be found in the book by Bruckheimer *et al.*, given above.

The book contains a number of examples which illustrate the application of algebraic concepts and a large number of exercises (with outline solutions). You should find the exercises valuable as an aid to revising the algebra component of the course material; in particular, they will help you to gain facility in the performance of standard algebraic manipulations.

Since the set book for our course is based upon material drawn from the Mathematics Foundation Course texts, those of you who are interested in taking your study of a particular concept further may find useful material in the relevant M100 course units. However, because of the integrated nature of the M100 material, the study of a particular text may necessitate background reading from a number of other texts (the extent of this interrelation can be determined by reference to the structural diagram which is at the front of each text). The set book for the M100 course is G. Polya, *How to Solve It*, Open University edition (Doubleday Anchor Books, 1970). This book presents a systematic approach to problem-solving in mathematics, which you will find of value if "making a start on a problem" is one of your difficulties.

1.0 INTRODUCTION

1.0.1 General Introduction to the Course

In producing this course the Course Team have been guided by the following aims.

(i) To provide an introductory course in mathematics for students whose primary interest lies in the study of topics from science and technology. The course aims to present a broad and modern account of the basic concepts of mathematics.

(ii) To present mathematics as one of the basic tools for developing techniques and solving problems in modern scientific and engineering disciplines.

(iii) To cover those areas of mathematics that are finding increasingly important applications in a modern technological society.

(iv) To impart mathematical skill in performing standard manipulations.

(v) To integrate the various subject areas described in the course into a cohesive description of mathematics as a subject, rather than as a loosely related collection of topics, thus providing a foundation for further study.

We feel sure that, by the time you have completed your study of the course material, you should have acquired a working knowledge of the basic concepts of mathematics, and the skill to apply mathematical techniques to a range of problems drawn from your other areas of study.

1.0.2 Presentation of the Course Material

The reading material of the course is presented to you in two forms.

On the one hand you are asked to study portions of the set book, *An Introduction to Calculus and Algebra*, which introduces the fundamental concepts of the subject under discussion. The set book includes examples and exercises which illustrate these concepts and *introduce* basic manipulative skills.

On the other hand, you are asked to study these correspondence texts which are concerned with guiding, by means of additional notes, examples and exercises, your study of the mathematical concepts treated in the set book. The texts also provide additional examples and discuss the development of mathematical techniques which are particularly relevant to students studying science and technology. Some of the problems presented in these texts are more demanding than the exercises associated with the material in the set book, and are designed to provide you with practice in solving problems. Such practice is of fundamental importance. A grasp of the theory and of basic manipulative skills, which is not accompanied by experience with problem solving, will be of limited value when you come to apply your mathematical knowledge to solving problems originating in other fields of study.

A number of case studies form an integral part of the course; each of these is introduced in a television programme and followed up in later programmes and in the correspondence texts. In these case studies we aim to relate mathematics to particular physical situations and to show how we can "model" the real world in terms of mathematics. We shall refer to these case studies as the relevant mathematical topics arise in the course. Thus, we shall demonstrate "mathematics in action" and highlight the importance of certain concepts.

Obviously in a half-credit course we cannot cover all the mathematics you are likely to require in solving the problems you will meet in your own subject area. Therefore, while we have to concentrate on those concepts which scientists and technologists apply, we also aim to cover important basic concepts, so that when you have to tackle a problem involving mathematics you will have some idea of which techniques can be applied to its solution.

The course consists of 17 units, each intended as one week's work. Each unit will include some or all of the following:

(i) a correspondence text which contains a guide to studying a portion of the set book, and self-assessment questions aimed at helping you to check your own progress;

(ii) a television programme lasting 25 minutes, backed up by television broadcast notes, aimed at showing you how mathematical concepts are developed and applied;

(iii) an assignment which you are required to complete and return for grading;

(iv) a radio programme lasting 20 minutes, backed up by radio broadcast notes aimed at guiding your study of particular concepts and dealing with difficulties that arise;

(v) a computing session based on a separate course text and aimed at enabling you to achieve basic skill at computer programming.

As a guide to the allocation of your study time, we suggest that you spend

5% of your time watching the *television programme* and working on any preparatory or supplementary material associated with it:

15% of your time working on the *assignment*.

5% of your time (when appropriate) listening to the *radio programme* and working on any material associated with it.

10% of your time (when appropriate) working on a *computing session* (about half of this time should be spent at a terminal if possible).

The remaining **65%–80%** of your time should be spent on working through the correspondence text and the associated parts of the set book.

You will probably have an opportunity to attend your study centre to meet other students and a class tutor from time to time. It is best to regard this activity as an addition to your normal work.

Each *text* will normally include the following components:

0 An INTRODUCTION to the week's work, including material which links the week's work to your previous studies.

1 CORE MATERIAL—covering:
a guide to the study of sections of the set book; each part of the guide will include a self-assessment question.

2 OPTIONAL MATERIAL—covering:
a guide to reading further sections of the set book;

examples of the application of mathematics to problems from science and technology;

discussion of a "theme problem" applying the concepts covered so far and problems for further study.

3 BACKGROUND MATERIAL—covering:
an explanatory presentation of concepts and techniques which are taken as prerequisites to the study of the week's unit, but which you may need to revise.

4 A SUMMARY of the week's work, including links with material to be studied subsequently.

Approximately *three-quarters of the course credit* for the course will be based on the *core material*; so we suggest that you spend at least three-quarters or more of the time which you devote to study of the text to this material.

Approximately *one quarter of the credit* will be based upon the *optional material* in the text *and* upon all other components of the course material (for example, computing sessions). You should always read through the optional material which refers to sections of the set book and note the terms and results given in these sections. There is no need to work through the exercises however, unless you wish to study the section in more detail. The optional material concerned with applications of mathematics to science and technology can be omitted entirely if you wish.

No credit rests directly upon the *background material*. It is provided solely to help you fill gaps in your existing knowledge when you come to study the core material. The background material can be omitted entirely if your mathematical knowledge is up to standard.

1.0.3   Introduction to This Week's Work

Mathematics is the language used to describe problems arising in most branches of science and technology. As with any other language, it is necessary to devote some time initially to learning the syntax (the study of the grammar of notational conventions) and the basic semantics (the study of the meaning of notational conventions). Consequently, much of the early units will be concerned with the language which we use to describe mathematical concepts and applications.

You will have already had *some* grounding in the language of mathematics. However we shall start "at the beginning". Our only assumptions are that you are familiar with the basic algebraic concepts of numbers, the arithmetic operations ($+$, $-$, $\times$, $\div$), the relations ($<$, $\leqslant$, $=$, $\geqslant$, $>$), variables, arithmetic expressions, equations and inequalities. Some of the more important aspects of these concepts are reviewed in this week's background material. If you feel that you need to review some of these concepts, then we suggest that you attempt the relevant exercise(s). If you cannot solve an exercise, then you should study the appropriate section of the background material. Our only other assumption is that you have a reasonable degree of facility with the manipulation and evaluation of algebraic expressions. If at the start this is difficult, then we suggest that you gain practice by writing out each algebraic manipulation given in the text yourself (if necessary expanding the argument by filling in any missing steps).

You may already have learnt a mathematical language for describing concepts which we develop in this course. For example, you may have met a notation for a function, which we develop in the first two units. You may well find that our notation is different from the one with which you are familiar. We have tried to develop the syntax and semantics of the mathematical language systematically side-by-side. It is most important that you master the language which we develop in these early units, because we use it throughout the course. However, it is also important that you relate your previous mathematical knowledge and experience to your present studies. We have provided supplementary notes to help you create this bridge between your past and present studies. These notes should not, however, be used as a means of building up "alternative ways of expressing the same concept"; they are only of value if they concern ideas and notations which you have encountered before. If the material is unfamiliar, then ignore it.

The *core material* this week starts by introducing the fundamental concepts of *set*, *mapping* and *function* and then develops the meaning and use of *graphs*. Graphs are important in elementary mathematics, for they give us a pictorial method of presenting and demonstrating a wide variety of mathematical concepts.

This material is then applied to a mathematical description of the concept of a *sequence* and that of a *limit of a sequence*. The limit concept is fundamental to the whole of our treatment of *calculus* and will appear often in subsequent units.

The *optional material* consists of three different parts:

(i) the use of the concepts of *set* and *mapping* to formally describe the concept of a *graph* and the solution of equations and inequalities;

(ii) examples of the use of the concepts of *mapping* and *sequence* to describe situations arising in problems in science and technology;

(iii) the application of the concept of a *sequence* to the preliminary analysis of methods for obtaining approximate solutions of equations in one variable.

When working on the core material, you should carefully study all the examples and exercises, and complete the self-assessment questions. However, when working the optional components you should pass over any unfamiliar concepts (making a note of them) in an attempt to gain a general impression of the material. You can then go over the more difficult areas again. Portions of the text designated "problems" can be regarded either as exercises or as examples, depending on your degree of mastery of the concepts involved.

1.1 CORE MATERIAL

1.1.1 The Concept of a Set

(a) Purpose

To define the term *set* and to introduce a notation for describing sets and relations between sets.

(b) Set Book

Read sections **1.1.0** and **1.1.1** and study section **1.1.2** (i.e. Volume 1, Chapter 1, section 2, pp. **5–9**) in the set book* and complete Exercise 1 at the end of the section.

(c) Notes

(1) The basic properties of the twelve sets of numbers referred to on page **6** are reviewed in section 1.3.1 of this text.

(2) When a set is defined by specifying some property possessed by every element of the set, then this property must be stated completely and unambiguously so that we can determine whether or not a particular object is an element of the set. Otherwise the set is not defined.

(d) Self-Assessment Questions

1   Write the following sets in standard notation.

   (i)   The set of all positive integers which are factors of the integer 12.
   (ii) The set of all real numbers whose cubes lie between 1 and 1000 inclusive.

2   List all the elements in the following sets.

   (i) $\{x: x \in Z^+, x < 20, x/3 \in Z^+\}$
   (ii) $\{(a, b): a \in Z^+, b \in Z^+, 3 \leqslant 2a + b \leqslant 5\}$.

3   In each of the following parts, determine the relation, if any, which exists between the sets A and B, and write it in standard notation.

   (i) $A = \{2, 4, 6, 8, 10\}$, $B = \{8, 4\}$.
   (ii) $A = \{x: x = 2n, n \in Z^+\}$, $B = \{n: n \in Z^+, n/2 \in Z^+\}$.
   (iii) $A = \{x, a, y\}$, $B = \{a, b, y\}$.
   (iv) $A = \{3, 5, 7, 11, 13\}$, $B = \{n: n \in Z^+, (n + 1)/2 \in Z^+\}$.

Solutions

In marking your answer remember that the elements of a set can be listed in *any* order.

1   (i)   Three possible solutions are

         $\{1, 2, 3, 4, 6, 12\}$, $\{x: x \in Z^+, 12/x \in Z^+\}$ and $\{x: x$ is a positive integer which is a factor of the integer 12$\}$.
   (ii) $\{x: x \in R, 1 \leqslant x^3 \leqslant 1000\}$

         Other expressions are possible.

2   (i) $\{3, 6, 9, 12, 15, 18\}$
   (ii) $\{(1, 1), (1, 2), (1, 3), (2, 1)\}$

3   (i) $B \subset A$
   (ii) $A = B$ (both are the set of even, positive integers).
   (iii) No relation.
   (iv) $A \subset B$ (B is the set of odd, positive integers).

---

* All set book section and page number references will be printed in **bold** figures to distinguish them from references to the course texts.

*Revision**

(1) If you got more than *one* part of questions 1 and 2 wrong, then revise the notation used to define a set (pp. **5–7**).

(2) If you got more than *one* part of question 3 wrong, then revise the concepts of a subset and of equality between sets (pp. **7–9**).

(e) Terminology

Defined in this section:

set
element (member) of a set
property defining a set
equality of sets
subset
proper subset
empty set

(f) Notation

Defined in this section:

$A, B, C, \ldots$	The names of sets.
$\{a, b, c, \ldots\}$	The set comprising elements $a, b, c, \ldots$
$a \in A$	a is an element of set A.
$\{x : x$ has property $P\}$	The set of *all* elements x having the given property P.
$A = B$	The sets A and B have the same elements.
$A \subseteq B$	The set A is a subset of the set B.
$A \subset B$	The set A is a proper subset of the set B.
$\varnothing$	The empty set.

(g) Additional Exercises (OPTIONAL MATERIAL)

1   Additional Exercise 1 (section **1.1.7, p. 32**).

This is a useful revision example. It tests knowledge of the notation used to express relations between sets and also knowledge of the properties of the twelve sets of numbers defined on page **6**. If you cannot follow the solution to this exercise, then you are advised to study the material in section 1.3.1.

2   If $A = \{x : x \in Z, x - 1 \leqslant 0\}$ and

$\qquad B = \{x : x \in Z, x + 2 \geqslant 0\}$,

list the elements of set C, the largest common subset of the sets A and B, and write a property definition of this set.

Solution 2

$$C = \{-2, -1, 0, 1\}.$$

A variety of descriptions may be given for C, the most obvious being

$$\{x : x \in Z, x - 1 \leqslant 0, x + 2 \geqslant 0\},$$
$$\{x : x \in Z, -2 \leqslant x \leqslant 1\}.$$

---

* We shall give revision notes occasionally in order to indicate how self-assessment questions should be used to guide your study. These revision notes are not included every time in order to avoid needless repetition.

1.1.2 The Concept of a Mapping

(a) Purpose

To define the term *mapping* and to introduce a notation for mappings.

(b) Set Book

Study section **1.1.3** (pp. **9–19**) and complete Exercises 1, 2 and 3.

(c) Notes

(1) A single element $a \in A$ can be mapped to a subset of B. That is, the mapping $A \longrightarrow B$ may include a correspondence of the form $a \longmapsto C$ where $a \in A$ and $C \subseteq B$. Example 3, page **10** illustrates a mapping which contains correspondences of this sort.

(2) The idea of one quantity depending upon another should be familiar to you. When the quantities are numbers (for example, two sets of measurements in an experiment), the relationship can often be expressed conveniently by a mathematical formula. For example, Hooke's Law, which states

"the extension in a wire is proportional to the force producing it",

can be expressed by the formula

$$e = kw,$$

where w is the force applied, e is the extension induced and k is an appropriate number. How is this formula written in the mapping notation?

The two sets involved are: W, the measured values of the weights applied—the DOMAIN—and E, the measured values of the induced extensions—the CODOMAIN. To complete the definition of a mapping, we need a rule that assigns to *each* value of $w \in W$ a corresponding value of $e \in E$ so that we can establish $f: w \longmapsto e$, $(w \in W)$.

The rule is a re-statement of Hooke's Law; that is, "to find the image of any $w \in W$ under f, multiply w by the constant k". Thus we could write Hooke's Law as

$$f: w \longmapsto kw \qquad (w \in W).$$

In this form we lose the name of the variable in the codomain; so a more appropriate notation to use in this case is

$$f: w \longmapsto e, \text{ where } f(w) = kw \qquad (w \in W),$$

which can be abbreviated to

$$f: w \longmapsto e = kw \qquad (w \in W).$$

That is, we express the image of w in two ways: as a variable e in the codomain and as a formula. Since in the case of a *function* each element in the domain has a unique image, we can write

$$e = f(w) = kw,$$

the familiar form for a physical law.

You will often find the variable in the domain referred to as the *independent variable* and the corresponding variable in the codomain referred to as the *dependent variable*.

(3) Mappings which map numbers to numbers, discussed on pages **17–21**, can be considered using ideas described in note (2) above and naming the variable in the codomain. For example, the mapping

$$f: x \longmapsto 6x^2 - 2x + 1 \qquad (x \in R)$$

can be written either as

$$f: x \longmapsto y, \text{ where } f(x) = 6x^2 - 2x + 1 \qquad (x \in R),$$

or as

$$f: x \longmapsto y = 6x^2 - 2x + 1 \qquad (x \in R).$$

In image form, we can abbreviate this formula to

$$y = f(x) = 6x^2 - 2x + 1 \qquad (x \in R),$$

the form with which you may be accustomed when describing a *functional relation* between variables x and y. When this form is used, it is normal to call x the *argument* of the function f.

The important point to remember is that, if we name the variable in the codomain, and if there exists an algebraic formula which defines the function (mapping), then we can write our function in image form. The notation introduced in section **1.1.3** enables us to specify mappings in which a more general relation is used to associate the elements in the domain with the elements in the codomain. In the set book we use only this one notation to describe functions. However, since the image form is in common use in science and technology, we shall use this notation freely elsewhere in the course material. Even when we use the image form, we shall continue to specify the domain and, where appropriate, the codomain of the function.

(d) Self-Assessment Questions

1 Let

$$A = \{1, 2, 4, 8, 16, 32, 64\}$$

and

$$B = \{0, 1, 2, 3, 4, 5, 6\}.$$

The mapping t from A to B is defined by the rule

"divide the element a from set A by 7 and take the remainder as the image of a".

Complete each of the following statements:

(i) $t: 16 \longmapsto ?$
(ii) $t: ? \longmapsto 3$
(iii) $t(32) = ?$
(iv) $t(A) = \{?\}$
(v)   Is it correct to write

$$t: A \longmapsto B?$$

2 Which of the following three options correctly complete the following statement?

If m is a mapping with domain A and codomain B, then

(i)   for every $a \in A$, $m(a)$ is an element of B.
(ii) B is the set of elements $\{b: b = m(a), a \in A\}$.
(iii) if $m: a_1 \longmapsto b_1$ and $m: a_2 \longmapsto b_2$ for each $a_1 \in A$ and $a_2 \in A$, and if $b_1 \neq b_2$ whenever $a_1 \neq a_2$, then m is a function.

3 f is the mapping

$$f: x \longmapsto \left(\frac{x+1}{x-1}\right) \qquad (x \in R, x \neq 1).$$

What is the image under f of

(i)   the set $\left\{-\dfrac{3}{2}, -\dfrac{1}{2}, 0, \dfrac{1}{2}, \dfrac{3}{2}\right\}$,

(ii) the set $\left\{x: -\dfrac{3}{2} \leqslant x \leqslant \dfrac{3}{2}\right\}$?

Solutions

1 (i)   2
  (ii) No element in A maps to 3.
  (iii) 4
  (iv) $\{1, 2, 4\}$
  (v)   No, because no element in A maps to 0, 3, 5 or 6. We could, however, write
        $t : A \longrightarrow B$.

2 (i)   True.
  (ii) False; B can contain elements other than the images of elements of A under m.
  (iii) False, the case $b_1 \neq b_2$ when $a_1 = a_2$ is not excluded; i.e. an element need not
        have a unique image.

3 (i) $\left\{\dfrac{1}{5}, -\dfrac{1}{3}, -1, -3, 5\right\}$

  (ii) This set has no image. It includes $x = 1$ so it is not a subset of the domain of the
        mapping f.

Revision

If you got more than one part of question 1 wrong then revise pages **9–14**.

If you got more than one part of questions 2 and 3 wrong then revise pages **14–19**.

(e) Terminology

Defined in this section:

mapping
image
domain
codomain
rule defining a mapping
function
variable
equality of mappings.

(f) Notation

Defined in this section:

$A \longrightarrow B$	Set A maps to set B.
$a \longmapsto b$	The element $a \in A$ maps to the element $b \in B$.
$f : A \longrightarrow B$	The mapping f maps set A to set B.
$f : a \longmapsto b$	The image of a under the mapping f is b.
$f(a)$	The image of a under the mapping f.
$f = g$	The mappings f and g have the same domain, codomain and rule.

(g) Additional Exercises (OPTIONAL MATERIAL)

1   Additional Exercise 2 (section **1.1.7**, p. **32**).

    This is a useful revision example for the material on pages **9–14**.

2   Additional Exercise 3 (pp. **32–33**).

    This is a useful revision example for the material on pages **14–19**.

1.1.3 The Concept of a Graph

(a) Purpose

To introduce the concept of a *graph* and to give practice in sketching graphs.

(b) Set Book

Study section **1.1.4** (pp. **20–25**) and complete Exercises 1, 2 and 3.

(c) Notes

(1) You have probably met graphs before. The process of constructing a graph of the mapping $f : A \longrightarrow R$ where $A \subseteq R$ is as follows: for each x in the domain of f we mark the points $(x, f(x))$ on the xy-plane. The set of all such points, that is, the subset of the plane defined by

$$\{(x, f(x)) : x \in A\}$$

is called the *graph* of f.

Graphs are a valuable aid to solving many problems; you must therefore become familiar with the techniques used for sketching a graph. It may help you to relate the material in section **1.1.4** to your previous knowledge, if you write the mappings in the image form described in section 1.1.2 (c), (3) of this text. Thus, in Example 1 on page **20**, if you label the axes x and y and write the function as either

$$g : x \longmapsto y, \quad \text{where } g(x) = x^2 - 2 \quad (x \in R)$$

or

$$g : x \longmapsto y = x^2 - 2 \qquad (x \in R),$$

then the graph is defined by the formula

$$y = g(x) = x^2 - 2,$$

which gives a value for y corresponding to each value of x.

(2) The notation $x \in [a, b]$, which is defined on page **24**, is important since the domain and codomain of many of the mappings and functions we deal with can be expressed in this form. We call the set $[a, b]$, where $a \leqslant b$, a *closed interval* because it includes the values a and b. (We call a and b the *end-points* of the interval.)

(d) Self-Assessment Questions

1 The mapping

$$m : x \longmapsto y \qquad (x \in [-2, 2])$$

is defined by the following graph.

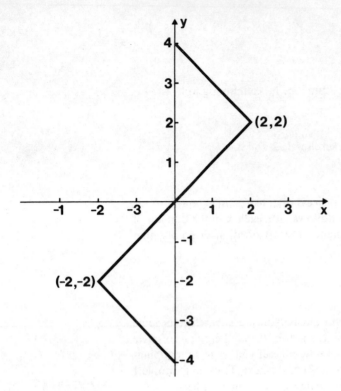

Is m a function?

What is the set of images of m?

2   Sketch a graph of the mapping m defined as follows:

$$m : x \longmapsto \begin{cases} -x - 2 & x \in [-2, -1] \\ x & x \in [-1, 2] \\ 4 - x & x \in [2, 4] \end{cases}$$

Is m a function?

What is the set of images of m?

3   Sketch the graph of the function

$$f : x \longmapsto \frac{|x|}{x} \qquad (x \in R, x \neq 0).$$

What is the set of images of f?

Solution

1   No; most values of x have a set of two values of y as their image; e.g. $f : 1 \mapsto \{1, 3\}$.
The set of images of m is the interval $[-4, 4]$.

2

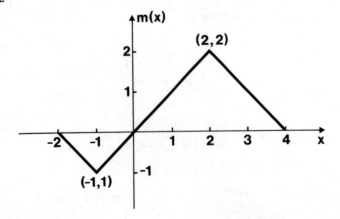

m is a function.

The set of images of m is the interval $[-1, 2]$.

3

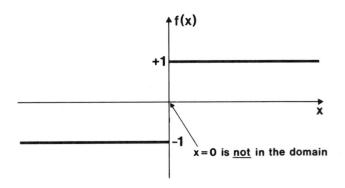

x = 0 is <u>not</u> in the domain

The set of images of f is the set $\{1, -1\}$.

(e) Terminology

Defined in this section:

graph
interval
modulus function

(f) Notation

Defined in this section:

$[a, b]$ The set of real numbers x such that $a \leqslant x \leqslant b$.
$|x|$ The modulus of x:

$$|x| = \begin{cases} x & \text{if } x \geqslant 0 \\ -x & \text{if } x < 0 \end{cases}.$$

(g) Additional Exercises (OPTIONAL MATERIAL)

1   Additional Exercise 4 (section **1.1.7**, p. **33**) provides additional practice in sketching graphs.

2   Consider the set of all functions

$$f : x \longmapsto ax + b \qquad (x \in R),$$

where a and b are any elements of the set $\{-2, -1, 0, 1, 2\}$. By plotting some of the 5×5 functions in this set, find a geometrical interpretation for a and b. (Note: f is called a *linear* function because the graph of f is a straight line.)

3   Consider the set of all functions of the form

$$f : x \longmapsto 1/(ax + b) \qquad (x \in R, \quad x \neq -b/a),$$

where a is any element of the set $\{-2, -1, 1, 2\}$ and b is any element of the set $\{-2, -1, 0, 1, 2\}$. By plotting some of the 5×4 functions in this set, find a geometrical interpretation for $-b/a$ and $1/b$, $(b \neq 0)$.

You may find it useful to repeat this exercise for the sets of functions of the forms

$$f : x \longmapsto ax^2 + bx + c \qquad (x \in R)$$
$$f : x \longmapsto 1/(ax^2 + bx + c) \qquad (x \in R, ax^2 + bx + c \neq 0).$$

Solution 2

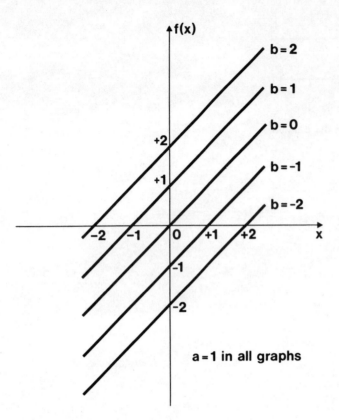

When the value of *a* is fixed (*a* = 1 in the figure above), the graphs of the functions are a set of parallel lines. The lines intersect the *y*-axis at a point distance *b* from the origin.

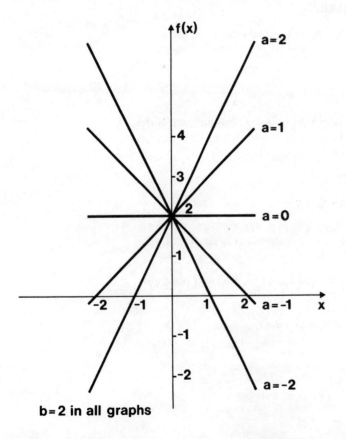

When the value of *b* is fixed (*b* = 2 in the figure above) the graphs of the functions are a set of straight lines of varying slope through the point on the *y*-axis at distance *b*

from the origin. If points $(x_1, f(x_1))$ and $(x_2, f(x_2))$ are two distinct points on the graph of f, then the slope of the graph of f is

$$\frac{f(x_2) - f(x_1)}{x_2 - x_1} = \frac{(ax_2 + b) - (ax_1 + b)}{x_2 - x_1} = a$$

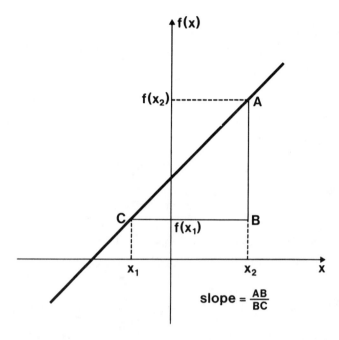

$$\text{slope} = \frac{AB}{BC}$$

Solution 3

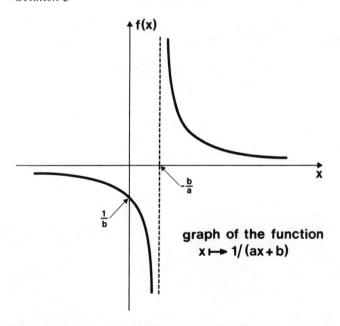

graph of the function
$$x \longmapsto 1/(ax + b)$$

The point $-b/a$ is the point at which f is not defined. To illustrate what is happening, we will consider the case when $a > 0$ and $b < 0$.

For $x > b/a$ $f(x)$ is positive and decreases from a very large positive value to a very small positive value (but never reaches zero), as x increases.

For $x < b/a$ $f(x)$ is negative and increases from a very large negative value to a very small negative value (but never reaches zero), as x decreases.

When $x = 0$ $f(0) = 1/b$, so that this is the point at which the graph crosses the $f(x)$-axis.

The cases when a and b are not as chosen above can be treated similarly.

1.1.4 The Concept of a Sequence

(a) Purpose

To define the term *sequence* and to introduce notation for describing sequences.

(b) Set Book

Read section **1.2.0** (p. **43**).

Study section **1.2.1** (pp. **43–47**) and complete Exercise 1.

Study section **1.2.2** (pp. **47–48**) and complete Exercises 1, 2 and 3.

(c) Notes

(1) The significance of the diagrammatic representation of a sequence defined by a recurrence formula (page **46**) should become clear when we consider computer programming later in the course. For the moment you can omit this material without affecting your understanding of the concept of a sequence.

(2) If you find determining the recurrence formula in Exercise 1 (page **47**) difficult don't worry about it. In many practical problems the form of the recurrence formula will be evident from the problem itself.

(3) There are two important sequences, which you may have met previously, namely the *arithmetic sequence* defined by

$$f : k \longmapsto a + (k - 1)b \qquad (k \in Z^+),$$

where a and b are real numbers, and the *geometric sequence* defined by

$$g : k \longmapsto ab^{k-1} \qquad (k \in Z^+),$$

where a and b are again real numbers. In many elementary mathematics texts these sequences are referred to as "progressions". We shall consider these sequences later in the course.

(4) Since a sequence can be specified by a function with domain Z^+ (or a subset of Z^+) and codomain the set of terms of the sequence, we can draw a graph of a sequence just as for any function. In this case, however, the graph will comprise a set of discrete points rather than a curve. Consider, for example, the sequence

$$f : k \longmapsto \frac{1}{k} \qquad (k \in Z^+).$$

The graph of the sequence has the form:

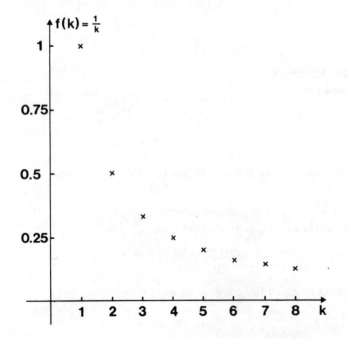

For an infinite sequence the graph extends indefinitely to the right.

(d) Self-Assessment Questions

1 (i) Express the sequence

$$1, 3, 7, 15, 31$$

as a function with domain $\{1, 2, 3, 4, 5\}$.

(ii) Evaluate the first four terms of the sequence defined by the recurrence relation

$$u_1 = 100$$

$$u_k = \left(1 + \frac{5}{100}\right)u_{k-1} \qquad (k = 2, \ldots, 10).$$

2 (i) Express the infinite sequence

$$x, -\frac{x^3}{3!}, \frac{x^5}{5!}, -\frac{x^7}{7!}, \ldots,$$

where x is a real number, using a function with domain Z^+.

The notation $n!$ means $1 \times 2 \times 3 \times \cdots \times n$.

(ii) Express the same sequence using a recurrence relation.

3 Draw a graph showing the first few terms of the infinite sequence

$$f : k \longmapsto \left(\frac{k-1}{k}\right) \qquad (k \in Z^+)$$

Solutions

1 (i) $f : k \longmapsto 2^k - 1 \qquad (k = 1, 2, \ldots, 5)$.
   (ii) 100, 105, 110.25, 115.7625

2 (i) $f : k \longmapsto (-1)^{k-1} \dfrac{x^{2k-1}}{(2k-1)!} \qquad (k \in Z^+)$

   (ii) $u_1 = x$

$$u_k = \frac{-x^2}{(2k-1)(2k-2)} u_{k-1} \qquad (k = 2, 3, \ldots)$$

3 $u_k = f(k)$

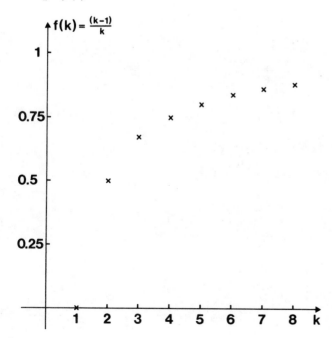

(e) Terminology

Defined in this section:

sequence
element (member or term) of a sequence
finite sequence
recurrence formula
infinite sequence

(f) Notation

Defined in this section:

a, b, c, d, e	The finite sequence with elements a, b, c, d and e.
$a, b, c, d, e, \ldots$	The infinite sequence whose first 5 members are a, b, c, d, e.
$1, 2, 3, \ldots, n$	The sequence whose elements are the natural numbers up to n.
u_k	The kth element of a sequence.

(g) Additional Exercises (OPTIONAL MATERIAL)

1 Additional Exercise 1 (section **1.2.4**, page **53**).

This provides an opportunity to revise the idea of describing a sequence in terms of a function with domain Z^+.

2 Plot graphs showing the first few terms of the infinite sequences defined by

$$f : k \longmapsto 1 + \frac{1}{k} \qquad (k \in Z^+);$$

$$g : k \longmapsto (-1)^k \frac{(1-k)}{(1+k)} \qquad (k \in Z^+).$$

Solution 2

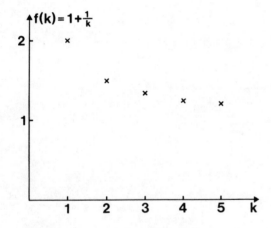

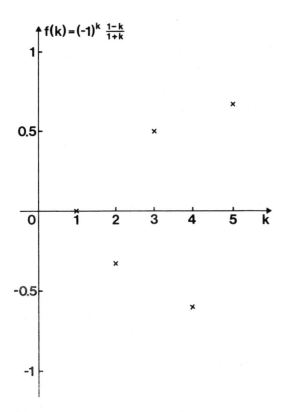

1.1.5 The Concept of a Limit

(a) Purpose

To introduce the concept of a limit of an infinite sequence, and to give practice in determining the limits of simple sequences.

(b) Set Book

Study section **1.2.3** (pp. **48–53**) and complete Exercises 1 and 2.

(c) Note

A sequence of successive approximations (see page **48**) can be defined as follows.

If the elements $u_1, u_2, u_3, \ldots$ of a sequence u are numbers which have as a limit the solution of an equation of the form $f(x) = 0$, then the sequence is called a sequence of *successive approximations* to the solution of the equation. The sequence given in the text,

$$0.3, 0.33, 0.333, 0.3333, \ldots$$

is a sequence of successive approximations to the solution of the equation $3x - 1 = 0$. It is of interest to see how this sequence is related to the equation. Suppose that we rewrite the equation in the form

$$9x = 3,$$

and then add x to each side to obtain

$$10x = 3 + x,$$

which can be rewritten as

$$x = (3 + x)/10.$$

Any x satisfying the original equation will clearly satisfy the new equation. But the new equation can be used to compute successive approximations to x starting from $x = 0$ using the equation as a recurrence relation. That is, the approximations to x are the elements of the sequence

$$u_1 = 0$$
$$u_k = (3 + u_{k-1})/10 \qquad (k = 2, 3, \ldots)$$

which is just the sequence quoted above. If lim u exists, then the intuitive definition tells us that, for k large, u_k (and hence u_{k-1}) are approximately equal to lim u. It is reasonable, therefore, to assert that lim u satisfies the equation

$$\lim u = (3 + \lim u)/10,$$

that is, lim u is a solution of the original equation. (Hence the assertion that lim $u = 1/3$.)

(d) Self-Assessment Questions

1 Classify the following statements as *true* or *false*.

   (i) If a sequence has limit L, no element of the sequence is equal to L.

  (ii) If all the elements of the convergent sequence $u_1, u_2, u_3, \ldots$ are positive then its limit must be positive.

 (iii) If a sequence has limit L then some of its elements must be less than $L + 0.0001$.

2 Which of the following sequences converge?

  (i) $1, -\dfrac{1}{2}, \dfrac{1}{3}, -\dfrac{1}{4}, \dfrac{1}{5}, -\dfrac{1}{6}, \dfrac{1}{7}, \ldots$

 (ii) $f : k \longmapsto (0.9)^k \qquad (k \in Z^+)$

(iii) $f : k \longmapsto (1 - (-1)^k) + \dfrac{1}{k} \qquad (k \in Z^+)$

Solutions

1 (i) False, for consider $f : k \longmapsto L \quad (k \in Z^+)$

  (ii) False, for consider $f : k \longmapsto \dfrac{1}{k} \quad (k \in Z^+)$

     (Zero is neither positive nor negative.)

 (iii) True, for k sufficiently large we can ensure u_k is as close to the limit L as we please.

2 (i) Converges to limit 0.
  (ii) Converges to limit 0.
 (iii) Diverges.

(e) Terminology

Defined in this section:

sequence of successive approximations
intuitive definition of a limit
convergent sequence
divergent sequence.

(f) Notation

Defined in this section:

u A sequence $u_1, u_2, u_3, \ldots$

lim u The limit of the sequence u.

(g) Additional Exercises (OPTIONAL MATERIAL)

1 Exercise 2, page 53 provides additional practice at finding the limits of sequences.

2 For each of the sequences given below find the smallest real numbers m and n ($m \leqslant n$) such that each term u_k of the sequence is contained in the interval $[m, n]$.

  (i) $f : k \longmapsto \dfrac{k - 1}{k} \qquad\qquad (k \in Z^+)$

 (ii) $f : k \longmapsto (-1)^k \left(\dfrac{k + 1}{k}\right) \qquad (k \in Z^+)$

Solution 2

(i) The elements of the sequence are

$$0, \frac{1}{2}, \frac{2}{3}, \frac{3}{4}, \dots$$

so $m = 0$, $n = 1$.

(ii) The elements of the sequence are

$$-2, \frac{3}{2}, -\frac{4}{3}, \frac{5}{4}, \dots$$

so $m = -2$, $n = \frac{3}{2}$.

1.2 OPTIONAL MATERIAL

1.2.1 The Concept of an Ordered Pair

(a) Purpose

To define the terms *mapping* and *function* in terms of the concept of an *ordered pair*.

(b) Set Book

Study section **1.1.5** (pp. **26–30**) and complete Exercises 1 and 2.

(c) Notes

(1) By specifying a mapping as a list of ordered pairs, that is, as a subset of the set $P \times Q$, where P is the domain and Q the codomain of the mapping, we can generalize the concept of a graph. This way of looking at mappings will be useful later in the course.

(2) The function $x \longmapsto \sin x$ mentioned in Example 3 will be considered in more detail later in the course. If you have not met it before, then you can simply regard this graph as an example in the same sense as the three which follow.

(d) Terminology

Defined in this section:

ordered pair
Cartesian product

(e) Notation

Defined in this section:

(p, q) The ordered pair p, q.
$P \times Q$ The Cartesian product of the sets P and Q.

1.2.2 The Concept of a Solution Set

(a) Purpose

To demonstrate the use of set notation to describe the solution of equations and inequalities.

(b) Set Book

Study section **1.1.6** (pp. **30–32**) and complete Exercises 1 and 2 and Additional Exercise 5 (section **1.1.7** p. **33**).

(c) Notes

(1) The section does not introduce any additional concepts; it simply shows how set notation can be used to describe the solution of equations and inequalities.

(2) If you are not familiar with the concept of an inequality, then you should omit the material on inequalities. We consider this topic in more detail in section 1.3.1.

(3) If you are not familiar with the properties of the sine function then omit part (iii) of Exercise 2.

(d) Terminology

Defined in this section:

solution set
equation
inequality

1.2.3 Examples

Example 1   Measuring the Power of an Engine

(To demonstrate the concept of a function.)

When selecting an engine to power a particular device one obviously must know the power which the engine will develop. The manufacturer of an engine can measure the power which it develops by means of a device called a dynamometer; with the aid of mathematics, he can express the power developed as a function of the engine speed. There are many different types of dynamometer; but their detailed construction need not concern us since they all operate on the same principle shown schematically in the following diagram:

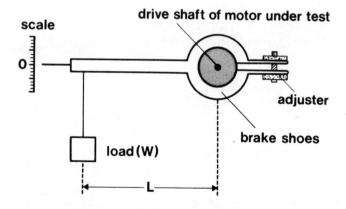

The brake shoes are tightened on to the drive shaft so as to supply resistance to rotation (the amount of resistance is controlled by the adjuster). For each adjuster setting there is a load (W) which will centre the arm on the scale. If the rate of revolution of the drive shaft of the motor (R) is measured by a revolution counter, then by varying the adjuster, a series of values of R, and the corresponding value of W needed to balance the arm, can be obtained. The functional relation between the variables R and W can be expressed either by tabulating or graphing the measured values of *R* and *W*. The result of a typical test would take the following form:

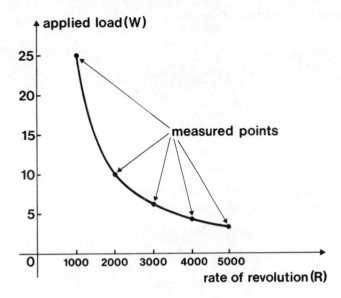

For the purpose of this example we need not concern ourselves either with the units in which the measurements are made, or even the accuracy of the measurements themselves. We are interested only in the fact that for each measured value of *R* there corresponds a measured value of *W*. That is, there is a *function* from the set of values of *R* to the set of values of *W*. By joining up the points in our graph we obtain an approximate function of the form

    rate of revolution $\longrightarrow$ applied load.

That is, for each value of *R* we can determine a corresponding value of *W*.

The power of an engine is defined in terms of the rate at which it can do work; work is defined as the product of the force applied to a body and the distance through which the body is moved by the force. Applying these definitions and the elementary ideas of mechanics we find the relation

    power $(P) = \alpha WR$,

where α is a constant dependent on the units in which *W* and *R* are measured and on the length of the dynamometer arm (L). Using this relation, and the measured functional relation between *W* and *R*, we can obtain a functional relation between either *P* and *R* or *R* and *W*. Consider the former case. Given a value of *R*, use the graph to determine the corresponding value of *W*, denoted by $W(R)$, then insert the values of *R* and $W(R)$ in the equation above to find the corresponding value of *P*. The result can be plotted as a graph of the function

    rate of revolution $\longrightarrow$ power developed

as shown below:

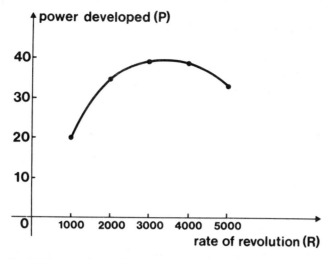

By joining up the points by a smooth curve we are able to use the graph to determine the value of P corresponding to each value of R.

This example demonstrates the two types of function which commonly arise in problems in science and technology, Firstly, the function

$$R \longrightarrow W$$

which is determined empirically by measurement and expressed as a graph or a table. Secondly, the functional relation between P, W and R which is expressed in terms of a formula $(P = \alpha WR)$ derived from the application of physical laws to a conceptual model of the physical situation. We have seen in section **1.1.2** how a formula can be expressed as a function using the " $\longmapsto$ " notation. In this case we would write

$$W \longmapsto P = \alpha WR,$$

where W is any element of the domain (the set of measured loads).

Example 2   Operation of a Remote Pumping Engine

(To demonstrate the concept of a limit.)

When a remote pumping engine is left to run indefinitely without supervision it is important that it will not overheat. By means of a simple experiment, and the application of mathematics, the manufacturer can obtain a design for a system that can operate without overheating.

Suppose that the engine operates on a fixed cycle in which it is on for a period of time s and then off for a period of time t and then on again for s, and so on. An experiment can be set up to measure the *maximum* increase in temperature during the *on* period; denote this temperature increase by I. If at the start of the on-period the temperature is T then at the end of the on-period it will be less than or equal to $T + I$ (equality arises only under the most unfavourable cooling conditions).

An experiment can also be set up to measure the *minimum* decrease in temperature during the *off* period. Because of the fact that the rate at which a body cools depends on its temperature, the temperature achieved at the end of the off–period can be approximated as a fraction of the initial temperature. Denote the *maximum* value of this fraction by D, then after one cycle in which the initial temperature is T the final temperature will be less than or equal to $D(T + I)$, where $D \in [0, 1]$ (equality arises only under the most unfavourable cooling conditions).

We now consider a sequence of operating cycles as shown in the following diagram.

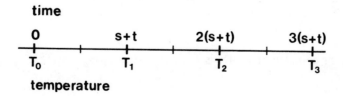

Then the temperature at each interval is given as follows:

$$t = 0, \qquad T = T_0$$
$$t = s + t, \qquad T = T_1 \leqslant D(T_0 + I)$$
$$t = 2(s + t), \qquad T = T_2 \leqslant D(T_1 + I)$$

and so on.

Consider the sequence

$$k \longmapsto T_k$$

defined by the recurrence relation:

T_0 is the initial value

$$T_{i+1} = D(T_i + I) \qquad (i = 0, 1, 2, 3, \ldots)$$

Clearly the limit of this sequence is greater than or equal to the eventual operating temperature of the engine at the start of each cycle. If this limit is denoted by L, then the engine will not overheat if $L + I$ (the maximum temperature achieved in the next on-period) is less than or equal to the maximum operating temperature.

Can we find the limit of this sequence?

Take the first few terms:

$$T_1 = D(T_0 + I)$$
$$T_2 = D(T_1 + I) = D(D(T_0 + I) + I)$$

i.e. $\qquad T_2 = D^2 T_0 + DI(D + 1)$

$$T_3 = D(T_2 + I) = D((D^2 T_0 + DI(D + 1)) + I)$$

i.e. $\qquad T_3 = D^3 T_0 + DI(D^2 + D + 1)$

The pattern is now becoming clear. We can write

$$T_{i+1} = D^{i+1} T_0 + DI(1 + D + D^2 + \cdots + D^i)$$

Since $D < 1$, we know the limit of the sequence

$$f : i \longmapsto D^{i+1} \qquad (i \in Z^+)$$

is zero*; but what about the sequence

$$g : i \longmapsto (1 + D + D^2 + \cdots + D^i) \qquad (i \in Z^+)?$$

(You may recognize the geometric series here; the following discussion is a standard method in this topic.)

Suppose we write

$$S = 1 + D + D^2 + \cdots + D^i$$

then

$$DS = D + D^2 + D^3 + \cdots + D^{i+1}$$

and subtracting

$$S(1 - D) = 1 - D^{i+1}$$

so that

$$S = (1 - D^{i+1})/(1 - D), \qquad D \neq 1.$$

Thus g can be written

$$g : i \longmapsto \frac{1 - D^{i+1}}{1 - D} \qquad (i \in Z^+)$$

which has limit $1/(1 - D)$.

It follows that the limit of the complete sequence is

* This will be formally proved in *Unit 3*.

$$L = ID/(1 - D)$$

so that the condition that the engine will not ultimately overheat is expressed as

$$L + I = \frac{ID}{1 - D} + I$$

$$= \frac{ID + I(1 - D)}{1 - D}$$

$$= \frac{I}{1 - D} \leqslant \text{maximum operating temperature.}$$

This tells the manufacturer how to vary I (which depends on the rate at which the engine heats up) or D (which depends on the rate at which the engine cools down) in his design so as to achieve a satisfactory range of operational temperatures. It is interesting to note that the final operating temperature is in fact independent of the initial temperature (T_0).

This result tells us only about long term effects; how can we be sure of choosing I and D so that the maximum temperature attained in any on-period $(T_i + I)$ is less than the maximum operating temperature? One way to do this is to ensure that T_i is less than or equal to the limit L for all i.

From our analysis above

$$T_i = D^i T_0 + DI(1 + D + D^2 + \cdots + D^{i-1})$$

$$= D^i T_0 + DI \frac{(1 - D^i)}{1 - D}$$

$$= \frac{D^i T_0(1 - D) + DI(1 - D^i)}{1 - D}$$

$$= \frac{D^i(T_0 - DT_0 - DI)}{1 - D} + \frac{DI}{1 - D}$$

$$= D_i \left(T_0 - \frac{DI}{1 - D} \right) + L$$

so that $T_i \leqslant L$ for all i if $T_0 - \frac{DI}{1 - D}$ is negative; i.e. $T_0 < \frac{DI}{1 - D}$.

Consequently, to be sure of satisfactory operation the manufacturer must design his engine so that

$$\frac{ID}{1 - D} \geqslant \text{initial operating temperature}$$

$$\frac{I}{1 - D} \leqslant \text{maximum operating temperature.}$$

Provided that these inequalities are satisfied I and D can take any values. Conversely, given values of I and D obtained by experiment, the permissible operating conditions can be estimated. For example, if I is 30°C and D is $\frac{2}{3}$ then the engine must be built to operate at up to 90°C and the starting temperature must be less than 60°C in order to be sure of satisfactory operation.

1.2.4 Functional Iteration

Suppose that we wish to solve the equation $f(x) = 0$. For the majority of equations it is not possible to specify the solution set directly in terms of a simple formula, so we are obliged to seek approximate ways of solving the equation. The idea of a sequence of approximations to the solution (see section 1.1.5) provides the basis for one approximate method.

Let us consider a simple example. Consider the equation

$$f(x) = x^2 - p = 0 \qquad x \in R, p \in R^+.$$

Our aim is to rewrite the equation $f(x) = 0$ in the form $x = F(x)$, in such a way that the new equation can be used as the basis for a recurrence relation

$$x_k = F(x_{k-1}) \qquad (k = 2, 3, \ldots)$$

for generating a sequence of successive approximations to one of the solutions of $f(x) = 0$. Clearly we can write

$$f(x) = F(x) - x = 0$$

that is

$$F(x) = f(x) + x = x.$$

For the particular case which we are considering we have

$$F(x) = x^2 - p + x$$

We must now ask whether the recurrence relation

$$x_k = F(x_{k-1}) \qquad (k = 2, 3, \ldots)$$

with some appropriate initial value x_1, generates a convergent sequence. We shall consider this problem, a bit at a time, over the next few units. In our first look at this problem, we want to apply the ideas which we have been developing in this unit to make a preliminary analysis.

At the outset we cannot even be sure that the computational process is well-defined, for a particular step in the iteration process may lead to a value of x_k for which $F(x_k)$ is not defined. Let us start by assuming that f has as its domain some interval $I = [a, b]$. Consequently the function F, defined as $F : x \longrightarrow x + f(x)$, also has domain $[a, b]$. Further let us assume that the image set of F is also an interval, which we denote by $J = [c, d]$.

Problem 1*

Is the fact that F has domain $[a, b]$ and image set $[c, d]$ an adequate condition to ensure that the computational process for generating the sequence $x_k = F(x_{k-1})$ $(k = 2, 3, \ldots)$, where $x_1 \in [a, b]$, is well defined? If not can you suggest a condition on F that is adequate?

Solution 1

The condition on F is *not* adequate to ensure that the computational process is well defined. To see this let us examine the iteration process in detail. Given any $x_1 \in [a, b]$ we can evaluate the image of x_1 under F. We obtain a value $F(x_1) \in [c, d]$. For the computational process to be well defined we want to be able to use the value of $F(x_1)$, call it x_2, as the starting point for the next step in the process. That is, given x_2 we want to be able to evaluate the image of x_2 under F to obtain $F(x_2)$. For $F(x_2)$ to exist we must insist that $x_2 = F(x_1) \in [a, b]$. This condition is satisfied if $[c, d] \subseteq [a, b]$, that is if the image set of F is a subset of the domain of F.

In Problem 1 we have seen that the computational process of generating the sequence of successive approximations based upon

$$x_k = F(x_{k-1}) \qquad (k = 2, 3, \ldots)$$

---

* These problems can be treated as *either* an exercise or an example.

is well defined if the set of images of F is a subset of the domain of F; we require $F(x) \in [a, b]$ for all $x \in [a, b]$. Since the codomain of F is any set containing the image set of F as a subset, it is convenient to regard the interval $[a, b]$ as both the domain and the codomain of F.

Problem 2

Is the fact that F has domain and codomain $[a, b]$ an adequate condition to ensure that $f(x) = 0$ has a solution in the interval $[a, b]$? If not, can you suggest a condition on F that is adequate?

Solution 2

The condition is *not* adequate. The easiest way to see this to specify a function $f(x)$, for which $f(x) = 0$ has no solutions, but which satisfies the conditions so far imposed on $F(x)$. For example, suppose f is defined as

$$f : x \longmapsto \begin{cases} +\dfrac{x}{2} & 1 \leqslant x \leqslant 2 \\[2mm] -\dfrac{x}{2} & 2 < x \leqslant 3 \end{cases}$$

Then F is defined as

$$F : x \longmapsto \begin{cases} \dfrac{3x}{2} & 1 \leqslant x \leqslant 2 \\[2mm] \dfrac{x}{2} & 2 < x \leqslant 3 \end{cases}$$

The set of images of F is $[1, 3]$ excluding the point $x = 1$, so that the set of images is certainly a subset of the domain $[1, 3]$ and F satisfies the condition imposed in the solution to Problem 1. But $f(x) = 0$ does not have any solutions in the interval $[1, 3]$.

To see the form of the additional condition which is required, let us consider the graph of the function F a little more closely. Since the domain of F is $[a, b]$, $F(x)$ is defined for all $x \in [a, b]$ including the end–points, that is, the points a and b. Further, since $F(x) \in [a, b]$ for all $x \in [a, b]$, we can assert that $(a, F(a))$ lies somewhere on the straight line joining point (a, a) to point (a, b), while $(b, F(b))$ lies somewhere on the line joining points (b, a) and (b, b).

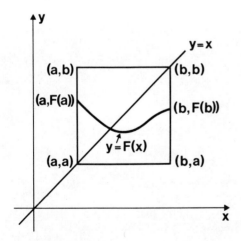

Remember that a solution of the equation $f(x) = 0$ also satisfies $F(x) = x$. The graph of F starts at $(a, F(a))$ and ends at $(b, F(b))$. If on its way it crosses the line with equation $y = x$ then at the point of intersection we have a value of x which is equal to $F(x)$. Intuitively the form of an additional condition is now clear; we require the function F to be such that, starting at the point $(a, F(a))$, it is possible to draw the graph of F and reach $(b, F(b))$ without lifting the pencil from the paper. This will ensure that we cross the line $y = x$ somewhere. We shall leave the mathematical interpretation of this condition until later.

The two conditions upon F, which we have derived in Problems 1 and 2, are adequate to ensure that the computational process is well-defined, and that $f(x) = 0$ has a solution within the interval $[a, b]$. These conditions do not exclude the possibility that $f(x) = 0$ has more than one solution in the interval $[a, b]$, nor have we established that the sequence generated by the recurrence relation $x_k = F(x_{k-1})$, $(k = 2, 3, \ldots)$, $x_1 \in [a, b]$, converges to any solution of $f(x) = 0$. We shall give a mathematical analysis of these problems later in the course; for the moment we return to our particular problem, the solution of $x^2 - p = 0$.

So far we have our recurrence relation in the form

$$x_k = F(x_{k-1}) = x_{k-1}^2 - p + x_{k-1}, \; x_1 \in [a, b], \; (k = 2, 3, \ldots).$$

We have not yet specified the domain. Let us specialize the problem by taking $p = 2$, then we know that formally the set of solutions is $\{-\sqrt{2}, +\sqrt{2}\}$.

Problem 3

Can we choose an interval $[a, b]$ so that the iteration $x_k = F(x_k) = x_{k-1}^2 - 2 + x_{k-1}$, $(k = 2, 3, \ldots)$ is well defined, and $[a, b]$ includes at least one element of the solution set of $f(x) = 0$?

Solution 3

(i) From Problem 1 we know that we must choose $[a, b]$ so that $F(x) \in [a, b]$ for all $x \in [a, b]$.

(ii) From Problem 2 we know that we should choose $[a, b]$ so that the graph of F in the interval $[a, b]$ can be drawn without lifting the pencil from the paper. To see whether this is possible let us plot the graph of the function $f : x \longmapsto x^2 + x - 2$.

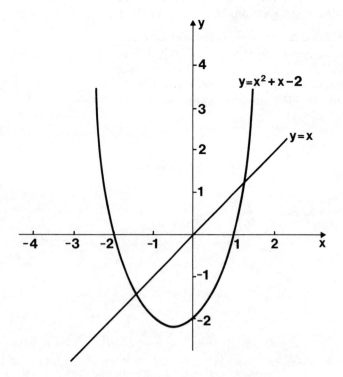

In this case there is nothing to stop us choosing the domain of F as R, but the choice $[-2\frac{1}{3}, 1\frac{1}{3}]$ also meets the above requirements and has the advantage of including just one of the two roots.

We now come to the main question: does the sequence which we can generate using F converge to $-\sqrt{2}$? With our present knowledge, we are not able to answer this question mathematically, so let us try an experiment. The table below shows the values obtained when we construct a sequence on particular choices for x_1.

x_1	-2	-1	0	1	-1.5	-0.5	$+0.5$
x_2	0	-2	-2	0	-1.25	-2.25	-1.25
x_3	-2	0	0	-2	-1.6875	0.8125	-1.6875
x_4	0	-2	-2	0	-0.8398	-0.5273	-0.8398
x_5	-2	0	0	-2	-2.1345	-2.2493	-2.1345
$\vdots$	$\vdots$	$\vdots$	$\vdots$	$\vdots$	$\vdots$	$\vdots$	$\vdots$
x_{22}	0	-2	-2	0	-0.7049	-0.5450	-0.7049
x_{23}	-2	0	0	-2	-2.2080	-2.2480	-2.2080
x_{24}	0	-2	-2	0	0.6673	0.8054	0.6673
x_{25}	-2	0	0	-2	-0.8874	-0.5460	-0.8874
$\vdots$	$\vdots$	$\vdots$	$\vdots$	$\vdots$	$\vdots$	$\vdots$	$\vdots$

On the basis of the intuitive definition of convergence which we developed in section 1.1.5 we would be inclined to the view that all of these sequences are divergent. Of course it is possible that $k = 25$ is not large enough, but since the sequences are oscillatory (they contain terms with different signs) there seems little likelihood of the sequences having a limit. In fact, we shall be able to show later in the course that F always generates a divergent sequence, whatever the starting value x_1, even if we start very close to the solution. For example, taking $x_1 = -1.4142$ (which is $-\sqrt{2}$ correct to four decimal places), we find that $x_{24} = 0.6279$, $x_{25} = -0.9779$, ..., which is the same sort of oscillatory behaviour as before.

Can we conclude from our result that it is not possible to solve the equation $x^2 - 2 = 0$ using functional iteration? Clearly we cannot, because we can rewrite the equation $f(x) = 0$ in many equivalent forms; each of these will give us a new form for the function F which may lead to a convergent sequence. For example, since we know $x = 0$ is not a solution of $f(x) = x^2 - 2 = 0$, we can exclude this point from the domain of F and write the equation in the form $f(x) = x - \dfrac{2}{x} = 0$ $(x \neq 0)$. The corresponding function F is then

$$F : x \longmapsto 2\left(x - \frac{1}{x}\right) \qquad (x \in R,\ x \neq 0).$$

Unfortunately this F also leads to divergent sequences. If, however, we rewrite $f(x) = 0$ as $\dfrac{1}{x} - \dfrac{2}{x} = 0$ then we have

$$F : x \longmapsto \frac{1}{x} + \frac{x}{2} \qquad (x \in R,\ x \neq 0)$$

which can lead to a convergent sequence.

Exercise 1

If F is defined as

$$F : x \longmapsto \frac{1}{x} + \frac{x}{2} \qquad (x \in R,\ x \neq 0),$$

using the graphical method of Problem 3, identify intervals $[a, b]$ which satisfy the conditions on F stated in the solutions to Problems 1 and 2.

Exercise 2

Select one of the finite intervals identified in Exercise 1 which includes the value $\sqrt{2}$ and show that the sequence generated by

$$x_k = F(x_{k-1}),\ (k = 2, 3, \ldots), \qquad x_1 \in [a, b],$$

converges to $\sqrt{2}$.

Exercise 3

Consider the sequence generated by

$$x_k = F(x_{k-1}) \qquad (k = 2, 3, \ldots), \qquad x_1 \in [1, 2]$$

$$F : x \longmapsto x[1 + (2 - x^2)/4] \qquad (x \in R, x \neq 0)$$

Show that the sequence is:

(i) a functional iteration for a solution of the equation $x^2 - 2 = 0$;

(ii) a sequence of successive approximations to $\sqrt{2}$.

Solution 1

The graph of F has the form

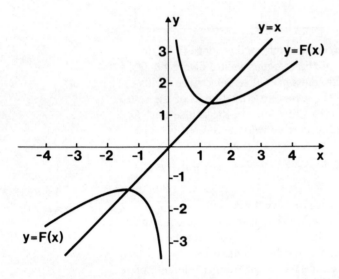

Since we require $F(x)$ to be defined for all $x \in [a, b]$, we must choose $[a, b]$ so that 0 is not included in the interval. Thus we must choose either $a > 0$ or $b < 0$. For $x > 0$, $F(x) > x$ for $x < \sqrt{2}$ and $\sqrt{2} < F(x) < x$ for $x > \sqrt{2}$. So, if we want the image set of $[a, b]$ to be a subset of $[a, b]$ we can choose $a < \sqrt{2}$ and $b \geqslant F(a)$. Thus a suitable choice of interval would be $[a, F(a)]$, where $0 < a < \sqrt{2}$; or, more generally $[a, b + F(a)]$. Similarly we can show that another set of intervals is $[a + F(b), b]$, where $-\sqrt{2} < b < 0$.

Solution 2

Choose $a = 1$, $b = 0$ so that the interval is $[1, 3/2]$. The convergence to $\sqrt{2}$ can be demonstrated by a sample computation; for example:

$$x_1 = 1$$

$$x_2 = \frac{3}{2}$$

$$x_3 = \frac{17}{12}$$

$$\vdots \qquad \vdots$$

Solution 3

(i) Let $f(x) = x^2 - 2$. Then we are seeking solutions of the equation $f(x) = 0$. Since $f(0) \neq 0$, the equation $\dfrac{x}{4} f(x) = 0$ has the same solution set, so we can redefine f as $f_1 : x \longmapsto x(2 - x^2)/4$ $(x \in R, x \neq 0)$. The corresponding function F is thus of the form required for functional iteration of the solution to $f_1(x) = 0$.

(ii) Plotting the graph of F in the interval $[1, 2]$, we see that, in the interval $[1, 2]$, F satisfies the conditions for the sequence generated by F to be well defined, and for the interval to contain a solution of $f(x) = 0$. The rest follows by analogy with Exercise 2.

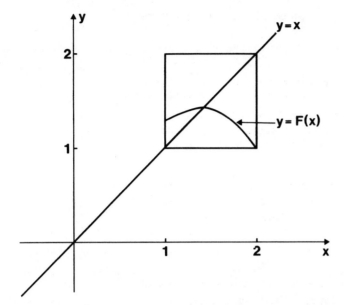

Summary

Given the equation $f(x) = 0$ with domain I, a subset of R, we define the function

$$F: x \longmapsto y = f(x) + x \qquad (x \in I)$$

If the function F is such that we can find an interval $[a, b] \subseteq I$ for which:

(i) for all $x \in [a, b]$, $F(x) \in [a, b]$
(ii) the graph of F in the interval $[a, b]$ can be drawn without lifting the pencil from the paper

then the sequence generated by

$$x_k = F(x_{k-1}), k = 2, 3, \ldots, \qquad x_1 \in [a, b]$$

is well defined and there is at least one value $s \in [a, b]$ for which $s = F(s)$, that is, s is a solution of $f(x) = 0$.

We have postponed the following problems for mathematical treatment later in the course:

(i) How do we ensure that $[a, b]$ contains only one element of the solution set of $f(x) = 0$?
(ii) How do we select a form for $f(x) = 0$ so that the sequence generated, using the corresponding form for F, is a sequence of successive approximations to the value s?

1.3 BACKGROUND MATERIAL

1.3.1 Properties of Numbers

(a) Purpose

To enable you to revise the properties of numbers required when solving problems arising in the course.

(b) Introduction

You learnt how to manipulate numbers at school; however, it is likely that it was not always made clear just what properties of a number you were using in carrying out each manipulation. When numbers are replaced by variables, which can take on the value of any of the members in a set of numbers, it is important that we understand the properties of the sets of numbers if we are to establish consistent algebraic manipulations on the variables we have defined.

Besides manipulating variables by means of arithmetic operations we are also interested in manipulating the relations which exist between variables (such as equality). Consequently we must also understand the relations which exist between the members of a particular set of numbers.

(c) The Integers and Rationals

The set of numbers which we use for counting is called a set of natural numbers, or the set of positive integers. If the arithmetic operation of addition is performed on two elements from the set of positive integers then the result is a positive integer. However things are not always as simple as this. Consider the outcome of the operations of subtraction shown below:

Operation	Example	Outcome
(i) subtraction	$11 - 9$	2: a positive integer
(ii) subtraction	$2 - 2$	not a positive integer
(iii) subtraction	$5 - 6$	not a positive integer

Example (ii)

$2 - 2$ represents "nothing"; we write the symbol 0 (zero or nought) to denote this null element. The most important property of zero is the fact that $a + 0$ is equal to a for all a in the set of positive integers.

Example (iii)

$5 - 6$ represents "1 less than 0" and this is denoted by writing -1. Notice that this is a second use of the ' $-$ ' sign. The most important property of ' $-$ ' in the second use is the fact that $-(-a)$ is equal to a.

To ensure that the operation of subtraction is defined and gives a result in the same set, we augment the set of positive integers by attaching 0 and the set of negative integers. The complete set is called the set of *integers*.

Notation

Z^+ denotes the set of positive integers
$\{1, 2, 3, 4, \ldots\}$
Z^- denotes the set of negative integers
$\{-1, -2, -3, -4, \ldots\}$
Z denotes the set of integers
$\{0, 1, -1, 2, -2, 3, -3, 4, -4, \ldots\}$

In Z, the operations of addition and subtraction are defined and the outcome is always in Z.

The set of integers is still inadequate for everyday practical arithmetic. Consider the outcome of the operation of division shown below.

Operation	Example	Outcome
(i) division	$18 \div 3$	6: in Z
(ii) division	$-3 \div 2$	Not in Z

The quotient of two elements in Z is not necessarily an element of Z. Note that the product of an element of Z with 0 gives 0, but that the division of an element of Z by zero is undefined. (It is meaningless.)

Rational Numbers

We augment the set of integers further. An element which can be represented in the form $\frac{p}{q}$, where

(i) $\frac{p}{q}$ means $p \div q$

(ii) p and q are integers

(iii) q is not zero (because $p \div 0$ is meaningless)

is called a *rational number*.*

Notice that $\frac{-3}{2}, \frac{3}{-2}, \frac{9}{-6}$, and so on, are all representations of the same rational number. It is conventional to choose the representation in which q is positive and where $\frac{p}{q}$ is in its lowest terms (that is, there exist no integers k, l, m such that $p = kl$ and $q = km$).

Thus, in the example above, we are concerned with the rational number $\frac{-3}{2}$. Other written representations of $\frac{-3}{2}$ are $-\left(\frac{3}{2}\right)$ and $-1\frac{1}{2}$.

Rational numbers can be manipulated by the rules for handling fractions which you learnt at school. The set of integers can be regarded as being contained in the set of rational numbers, because we can regard the integer a as a non-standard representation of the rational $\frac{a}{1}$.

Straight Line Representation

The rational numbers can be represented by points on a straight line as shown below:

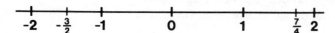

No matter how close the two rational numbers $\frac{m}{n}$ and $\frac{p}{q}$, there is always a rational number between them: the number $\frac{m}{n} + \frac{1}{2} \times \left(\frac{p}{q} - \frac{m}{n}\right)$.

this distance as small as we please

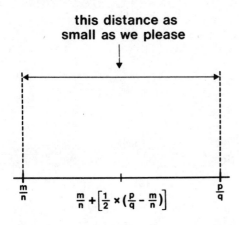

---

* Note that throughout this section we are not *defining* anything: we are merely describing the various types of numbers.

Notation

Q^+ denotes the set of positive rational numbers.

Q^- denotes the set of negative rational numbers.

Q denotes the set of positive and negative rational numbers and zero.

In Q the operations of addition, subtraction, multiplication and division by a non-zero number are defined and the outcome is always in Q. The set Q of rational numbers includes the set Z of integers.

Exercise

Identify the set of numbers (from the sets Z^+, Z^-, Z, Q^+, Q^-, and Q) in which each of the following expressions can be evaluated.

(i) $(5 - 9) \times (1 - 4)$

(ii) $(-1) \times (2\frac{1}{4} + \frac{3}{4})$

(iii) $(9 + 3)/4$

(iv) $\left(\dfrac{1}{2} - \dfrac{5 + 2}{14}\right) \times 3$

Solution

(i) Z or Q.

(ii) Q.

(iii) Z^+, Z, Q^+, Q.

(iv) Q.

(d) The Irrationals and Real Numbers

If we take a positive integer $r > 1$ and a positive integer $s < r$ then the rational $\dfrac{m}{n} + \left[\dfrac{s}{r}\left(\dfrac{p}{q} - \dfrac{m}{n}\right)\right]$ lies on the number line between the rationals $\dfrac{m}{n}$ and $\dfrac{p}{q}$. By varying r and s in $\dfrac{m}{n} + \dfrac{s}{r}\left(\dfrac{p}{q} - \dfrac{m}{n}\right)$, we are able to show that between any two rational numbers we can construct as many rational numbers as we please. Thus we might be tempted to conclude that every point on the line represents a rational number. This is not in fact the case: it can be shown (although we shall not do so) that between any pair of rationals there are other numbers which are not rationals.

We will show that the solution of $x^2 - 2$ is not a rational number; the method of proof is: ASSUME the solution is rational; hence, establish a contradiction of some known fact which enables one to REJECT the assumption.

Proof

ASSUME that the rational $\dfrac{p}{q}$, in its lowest terms, is a solution of $x^2 = 2$; then

$$\left(\frac{p}{q}\right)^2 = 2$$

i.e. $\qquad p^2 = 2q^2 \qquad\qquad\qquad\qquad$ (1)

so p^2 is even (divisible by 2).

Since p^2 is even, p is even, as the square of an odd number is odd. So

$$p = 2P \qquad\qquad\qquad\qquad (2)$$

where P is an integer.

Substituting (2) for p in (1), we obtain

$$(2P)^2 = 2q^2$$

i.e.

$$4P^2 = 2q^2$$

or

$$q^2 = 2P^2$$

so q^2 is even.

Since q^2 is even, q is even.

So $\dfrac{p}{q} = \dfrac{\text{even}}{\text{even}}$ is not in its lowest terms; this is a contradiction. Hence we can REJECT the assumption; in other words the solutions of $x^2 = 2$ is not a rational number.

Numbers like $\sqrt{2}$ (which is a solution of $x^2 = 2$), which cannot be represented by any integer fraction, are called *irrational numbers*. An important result, which we are not in a position to prove, is that the set of all rational and irrational numbers, called the set of real numbers, is "complete", that is, in some sense the real number system covers all points on a line leaving no "holes".

Notation

I^+ denotes the set of positive irrational numbers.
I^- denotes the set of negative irrational numbers.
I denotes the set of positive and negative irrational numbers and zero.
R^+ denotes the set of positive real numbers.
R^- denotes the set of negative real numbers.
R denotes the set of positive and negative real numbers and zero.

In R the operations of addition, subtraction, multiplication and division (by a non-zero number) are defined and the outcome is always in R.

Exercise

Use the fact that $\sqrt{2}$ is irrational to prove that if a is a rational number, then $a + \sqrt{2}$, $a - \sqrt{2}$ and $a/\sqrt{2}$ are also irrational. Will the result be true if $\sqrt{2}$ is replaced by any other irrational number?

Solution

Consider the expression $a + \sqrt{2}$. Suppose that $a + \sqrt{2}$ is rational and has value b; that is, $a + \sqrt{2} = b$ where a, b are members of Q. If we add $(-a)$ to both sides we obtain $\sqrt{2} = b + (-a)$. But if a, b are rational then $b + (-a)$ is a rational representation for $\sqrt{2}$. But this is a contradiction; so our assumption that $a + \sqrt{2}$ is rational must be false. The other proofs follow a similar pattern.

The results would hold for any irrational number; in the above argument, we used only the fact that $\sqrt{2}$ is irrational.

(e) Decimal Representation of Numbers

The use of a decimal representation for numbers occurs in most arithmetic work. It involves expressing a number as a combination of a whole number (integral) part and a fractional part in a particular way. Each part is represented as a sum of powers of ten (positive or zero integer powers for representing the integral part, and negative integer powers for the fractional part). A decimal number, which is written

$$a_m a_{m-1} \cdots a_1 a_0 \cdot b_1 b_2 \cdots b_n,$$

where the as and bs are decimal digits (i.e., digits from the set $0, 1, 2, \ldots 9$) means

$$a_m(10^m) + a_{m-1}(10^{m-1}) + \cdots + a_1(10^1) + a_0 + \frac{b_1}{10} + \frac{b_2}{10^2} + \cdots + \frac{b_n}{10^n}.$$

Decimal Representation of Rational Numbers

Every rational number of the form $\dfrac{p}{q}$ can be "divided out" to give a decimal representation of the rational number. For example,

$$\frac{14}{11} = \frac{11+3}{11} = 1 + \frac{3}{11} = 1 + \frac{1}{10}\left(\frac{30}{11}\right) = 1 + \frac{1}{10}\left(2 + \frac{8}{11}\right)$$

$$= 1 + \frac{2}{10} + \frac{1}{10}\left(\frac{1}{10}\left(\frac{80}{11}\right)\right)$$

$$= 1 + \frac{2}{10} + \frac{7}{100} + \frac{1}{100}\left(\frac{3}{11}\right) \text{ and so on.}$$

So that

(i) $\dfrac{14}{11} = 1.2727(\dot{2}\dot{7})$

where the dot over the digits in parentheses indicates that digits included in parentheses are repeated indefinitely.

(ii) $\dfrac{7}{5} = 1.40(\dot{0})$

(iii) $\dfrac{22}{7} = 3.142857(\dot{1}4285\dot{7})$

(iv) $\dfrac{1}{9} = 0.1111(\dot{1})$

It can be shown that if the decimal representation of a number

(i) terminates with a string of zeros as in example (ii),
(ii) possesses a recurring pattern as in examples (i), (iii) and (iv), in which a non-zero number or group of numbers is repeated indefinitely, then it is a rational number.

Decimal Representation of Irrational Numbers

It can be shown that a non-terminating decimal representation, which does not have a recurring pattern as described above, is not a rational number. Such representations must represent irrational numbers.

Exercise

Consider the number α whose decimal representation is $\alpha = 0.101001000100001\ldots$ that is a sequence of 1 s each separated by a sequence of 1, 2, 3, ... zeros. Is α a rational number?

Solution

α is not a rational number because its decimal representation neither terminates in a sequence of zeros nor repeats a sequence of digits. The fact that the decimal representation of α follows a regular pattern is of no significance.

(f) Inequalities

Sets of numbers possess another property, the fact that they can be *ordered*. The fact that integers, rationals, irrationals and consequently reals can be arranged according to size is a valuable property and one which we have already used to allow us to construct a representation of number systems in terms of points on a line. Ordering is achieved by using the concept "is greater than" (*a* is greater than *b* is denoted by $a > b$).

The basic properties of the relation "is greater than" can be summarized as follows;

(i) For any (integer, rational, irrational, real) number *a* exactly *one* of the following statements is true:

(a) *a* is greater than 0
(b) *a* is equal to 0
(c) $-a$ is greater than 0

(ii) If a and b are any pair of positive (integer, rational, irrational, real) numbers then $a + b$ is positive and ab is positive.

Notation

$a < b$ denotes "a is less than b".

$a > b$ denotes "a is greater than b".

$a \leqslant b$ denotes "a is less than or equal to b".

$a \geqslant b$ denotes "a is greater than or equal to b".

A statement such as $a > b$ is called an *inequality*. You will often need to manipulate inequalities, consequently we list below some of their more important properties.

   (i)   If $a(\geqslant)b$, then $a + c(\geqslant)b + c$*.

   (ii)   If $a(\geqslant)b$ and $k > 0$, then $ka(\geqslant)kb$.

   (iii)   If $a(\geqslant)b$ and $b > 0$, then $\dfrac{1}{a}(\leqslant)\dfrac{1}{b}$.

   (iv)   If $a(\geqslant)b$ and $k < 0$, then $ka(\leqslant)kb$.

It is not difficult to demonstrate these results. For example, $a < b$ implies $b - a < 0$, so that $(b + c) - (a + c) < 0$ which implies $a + c < b + c$ thus establishing half of property (i). You can establish similar properties for combining inequalities. For example

   (v)   If $a(\leqslant)b$ and $c(\leqslant)d$, then $a + c(\leqslant)b + d$.

   (vi)   If $a > b > 0$ and $c > d > 0$, then $ac > bd$.

Similar properties can be written down for $\leqslant$ and $\geqslant$ by allowing for the effect of the case of equality. For example, from (v) we can obtain:

   If $a > b$ and $c \geqslant d$, then $a + c > b + d$.

   If $a \geqslant b$ and $c \geqslant d$, then $a + c \geqslant b + d$.

   If $a \geqslant b$ and $c > d$, then $a + c > b + d$.

   and so on.

You should manipulate inequalities with care. It is often useful to make use of the following graphical representation of equations and inequalities.

In the following diagram the equation $x^2 + y^2 = 1$ together with the inequalities $x^2 + y^2 > 1$ and $x^2 + y^2 < 1$ divide the plane into three parts.

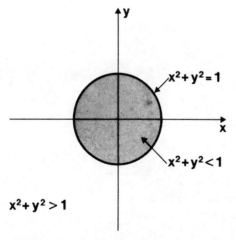

Similarly for the equation $y = x + 1$ together with the inequalities $y > x + 1$ and $y < x + 1$.

---

* This is an abbreviated way of writing the two statements:

   if $a > b$, then $a + c > b + c$

and

   if $a < b$, then $a + c < b + c$

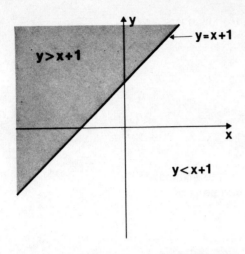

In the following diagram the shaded region represents that part of the plane in which $x^2 + y^2 < 1$ and at the same time $y > x + 1$.

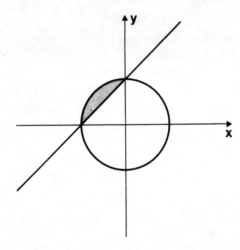

We shall often use graphical methods for selecting sets of values for which inequalities are satisfied; you should examine each use of this technique until you become familiar with it.

Exercise 1

Demonstrate the truth of the statement:

if $a > b$ then $-a < -b$.

Exercise 2

Find the values of x for which

$$x^2 - 3x - 4 \leqslant 0.$$

Solution 1

$a > b$ implies that $a - b > 0$.

If we multiply each side of this latter inequality by (-1) and use result (iv) above we have $(-1) \times (a - b) < (-1) \times 0 = 0$. This is $-a + b < 0$, which can be written $-a - (-b) < 0$, which implies $-a < -b$.

Solution 2

Two approaches are possible.

(i) Use the graphical method; if we plot the curve $y = x^2 - 3x - 4$, we obtain

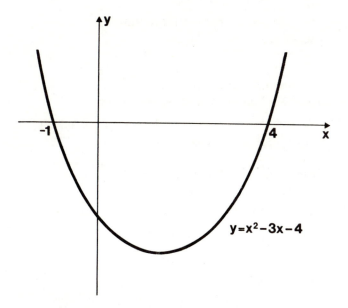

$y = x^2 - 3x - 4$

We see that $x^2 - 3x - 4 \leqslant 0$ for $-1 \leqslant x \leqslant 4$.

(ii) Factorize the expression to obtain $(x - 4)(x + 1)$ and put $a = x - 4$, $b = x + 1$. Then note that $ab \leqslant 0$ if and only if

either $a \leqslant 0$, $b \geqslant 0$, that is, $x \leqslant 4$ and $x \geqslant -1$

or $a \geqslant 0$, $b \leqslant 0$, that is, $x \geqslant 4$ and $x \leqslant -1$

Since the latter conditions cannot be satisfied simultaneously the solution is $\{x : x \in R, \ -1 \leqslant x \leqslant 4\}$.

1.3.2   Solution of Equations

(a) Purpose

To enable you to revise the methods used to solve certain simple types of equation.

(b) Introduction

A variable is a quantity which can take the value of any of the numbers in a set of numbers. For example, we might define a variable whose value was any real number greater than 1 but less than 2. In what follows we shall assume that the variables we encounter can take as their values any real numbers unless otherwise stated.

A variable is usually represented by a letter; variables and numbers can be combined by means of the arithmetic operations ($=$, $-$, $\times$, $\div$) and parentheses to make arithmetic expressions. For example, x, $x + 7$, $x \times y$, $(x + a) \div (y + b)$ are all arithmetic expressions. We shall assume that you know how to manipulate simple expressions. We shall also assume that you know how to evaluate an expression given the values of the variables it contains.

An equation is a statement that two expressions are equal. The following, in which x and y are variables, are all equations

$$x + 1 = 7, \quad x^2 = y - 1, \quad x^2 - 5 = 0, \quad \frac{1}{x} = y + 2.$$

To solve an equation we determine particular values for each of the variables in the equation which make the statement implied by the equation true. For example the equation $x + 1 = 7$ is satisfied (that is, it is a true statement) if and only if x has value 6. The values which satisfy an equation are called its *solution*. There are a number of *allowable operations* (see below) which can be performed on equations to give a new equation with the same solution as the original equation. Equations derived by performing allowable operations on another equation are called *equivalent equations*.

How do we solve an equation? Well we perform allowable operations to derive equivalent equations, until we obtain an equation (or a set of equations) of the form

"variable" = "expression"

in which the "expression" contains only numbers or variables whose values are assumed to be already known (such variables are often called constants). We will now consider how to solve certain simple equations.

(c) Linear Equations in One Variable

A linear equation in one variable is an equation which can be reduced by allowable operations to the form $ax + b = 0$, where a and b are constants and $a \neq 0$. Consequently we consider how to determine the solution of the equation

$$ax + b = 0, \tag{1}$$

where a and b are constants.

Subtracting b from both sides produces an equivalent equation

$$ax = -b. \tag{2}$$

Since $a \neq 0$, we can *divide both sides of* (2) *by a*, giving another equivalent equation

$$x = \frac{-b}{a} \tag{3}$$

(3) is the solution of (1) provided $a \neq 0$. In the above, the two allowable operations used have been printed in italic type. Note that a solution can be checked by substituting it into the original equation and seeing if the left-hand side (LHS) and the right-hand side (RHS) are, in fact, equal. Substituting (3) in (1) gives

$$\text{LHS} = a\left(\frac{-b}{a}\right) + b = 0 = \text{RHS}.$$

Graphical Interpretation

The values of x and y which satisfy the equation $y = ax + b$ are the co-ordinates of a set of points which lie on a straight line when plotted as a graph. The problem

solve $ax + b = 0$

can be represented geometrically by

find the intersection of the straight line whose equation is $y = ax + b$ and the x-axis (for on the x-axis $y = 0$).

The x-co-ordinate of their point of intersection is the solution of $ax + b = 0$.

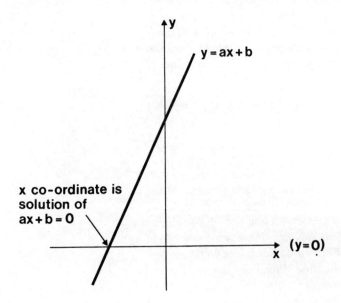

Exercise

Solve the equation $2x - 3 = 5$.

Solution

Subtracting 5 from both sides gives $2x - 8 = 0$. Using the result in Equation (3) gives $x = -(-8)/2 = 4$.

(d) Quadratic Equations in One Variable

A quadratic equation in one variable is an equation which can be reduced by allowable operations to the form $ax^2 + bx + c = 0$, where x is the variable and a, b, c are constants and $a \neq 0$. Consequently we consider how to determine the solution of the equation

$$ax^2 + bx + c = 0 \tag{1}$$

where a, b and c are constants and $a \neq 0$.

(If a were 0, the quadratic equation would reduce to a linear one whose solution we have just considered.) The solutions of (1), if they exist, are given by a pair of equations

$$x = \frac{-b + \sqrt{b^2 - 4ac}}{2a}$$
$$x = \frac{-b - \sqrt{b^2 - 4ac}}{2a} \tag{2}$$

Since (2) involves the operation " take the square root ", it gives real number solutions of (1) only when $b^2 - 4ac$ is zero or positive. The derivation of (2) is given below:

Subtracting c from both sides of (1)

produces an equivalent equation

$$ax^2 + bx = -c \tag{3}$$

Adding $\dfrac{b^2}{4a}$ to both sides of (3)

produces an equivalent equation

$$ax^2 + bx + \frac{b^2}{4a} = \frac{b^2}{4a} - c$$

which can be written in the form

$$a\left(x + \frac{b}{2a}\right)^2 = \frac{b^2 - 4ac}{4a} \tag{4}$$

Dividing each side of (4) by a

produces an equivalent equation

$$\left(x + \frac{b}{2a}\right)^2 = \frac{b^2 - 4ac}{4a^2} \tag{5}$$

Taking the square root of each side of Equation (5)

produces a pair of equations (because $x^2 = (-x)^2$)

$$x + \frac{b}{2a} = \sqrt{\frac{b^2 - 4ac}{4a^2}} \tag{6}$$

$$x + \frac{b}{2a} = -\sqrt{\frac{b^2 - 4ac}{4a^2}} \tag{7}$$

Subtracting $\dfrac{b}{2a}$ from each side of (6) and (7)

produces a pair of equations

$$x = \frac{-b - \sqrt{b^2 - 4ac}}{2a}$$

$$x = \frac{-b + \sqrt{b^2 - 4ac}}{2a}$$

Exercise

(i) Solve $y^2 + 2y - 3 = 0$

(ii) Solve $4x^2 + 4x + 1 = 0$

(iii) Solve $2x^2 - 3x + 10 = 0$

Solution

(i) By direct substitution in (2) we get

$y = -3$ or 1.

(ii) $x = -\frac{1}{2}$ or $-\frac{1}{2}$.

Note that in this case $b^2 - 4ac = 0$; we say that the equation has two identical (or coincident) solutions.

(iii) This time, substituting in (2) gives

$$x = \frac{3 + \sqrt{-71}}{4} \quad \text{or} \quad \frac{3 - \sqrt{-71}}{4}$$

In this case $b^2 - 4ac = -71$; the equation has no real number solutions.

Geometrical Interpretation

$ax^2 + bx + c = 0$ has solutions

$$x = \frac{-b + \sqrt{b^2 - 4ac}}{2a} \quad \text{and} \quad x = \frac{-b - \sqrt{b^2 - 4ac}}{2a}$$

provided $a \neq 0$ and $b^2 - 4ac$ is zero or positive. Since the values of x and y satisfying $y = ax^2 + bx + c$ represent a curve when plotted as a graph and we require to find the values of x for which $ax^2 + bx + c = 0$, the problem

solve $ax^2 + bx + c = 0$

can be represented geometrically by

find the intersections of the curve $y = ax^2 + bx + c$ and the x-axis.

The x co-ordinates of their points of intersection are the solutions of $ax^2 + bx + c = 0$. In the next three figures we have taken a positive, since $-ax^2 - bx - c = 0$ is an equivalent equation to $ax^2 + bx + c = 0$. In the following figure $b^2 - 4ac$ is positive; the equation has two distinct solutions.

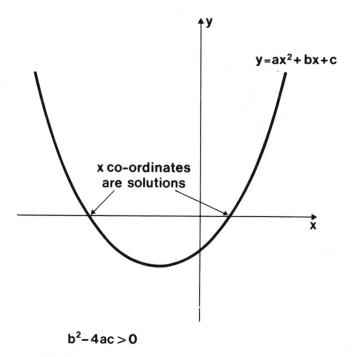

$b^2 - 4ac > 0$

In the next figure $b^2 - 4ac = 0$; the curve touches the x-axis and the equation has two identical solutions.

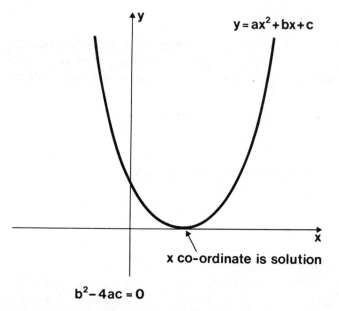

$b^2 - 4ac = 0$

In the next figure $b^2 - 4ac$ is negative; the curve and the x-axis do not intersect or touch; the equation has no real solutions.

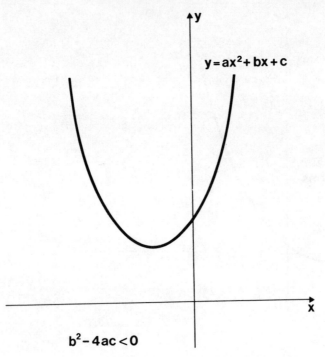

$b^2 - 4ac < 0$

This geometrical interpretation shows that a quadratic equation has either two distinct solutions or two identical solutions or no solutions.

(e) Linear Equations in Two Variables

A linear equation in two variables is an equation which can be reduced by allowable operations to the form $ax + by + c = 0$, where x and y are variables and where a, b and and c are constants $a \neq 0$ and $b \neq 0$. A solution of such an equation is a pair of values, one for x and one for y.

Geometrical Interpretation

How can we represent the set of all solutions of this equation?

If we subtract $ax + c$ from each side and then divide each side by b

we produce the equivalent equation

$$y = -\frac{a}{b}x - \frac{c}{b}$$

whose solution set is represented by a straight line.

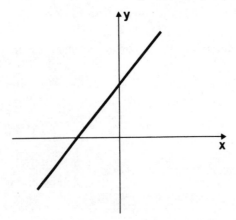

Any point on this line is a solution of $ax + by + c = 0$.

Exercise 1

By considering the geometrical interpretation of the equations

$$a_1 x + b_1 y + c_1 = 0 \qquad (1)$$

$$a_2 x + b_2 y + c_2 = 0, \tag{2}$$

state whether you think that there is an element of the set of all solutions of (1) which also belongs to the set of all solutions of (2).

Solution 1

The sets of solutions of $a_1 x + b_1 y + c_1 = 0$ and $a_2 x + b_2 y + c_2 = 0$ represent straight lines which will either

   (i) intersect   (ii) be coincident   (iii) be parallel

In case (i) there is exactly one pair of values of x and y satisfying both (1) and (2); in case (ii) there are infinitely many; in case (iii) there are none.

Two Simultaneous Linear Equations in Two Unknowns

We now consider how to determine the solution of a pair of linear equations in the form

$$a_1 x + b_1 y + c_1 = 0 \tag{1}$$

$$a_2 x + b_2 y + c_2 = 0 \tag{2}$$

where x and y are variables and a_1, b_1, c_1, a_2, b_2, c_2 are constants $a_1 \neq 0$, $b_1 \neq 0$, $a_2 \neq 0$, $b_2 \neq 0$. We saw in the answer to the last exercise that (1) and (2) have a unique solution unless the lines they represent are parallel or coincident. We will devise an algebraic representation for this condition below. Equations (1) and (2) can be solved by the "method of elimination". To eliminate x and obtain a linear equation in y alone, *we multiply equation (1) by a_2 and equation (2) by a_1 to produce the equivalent equations*

$$a_2 a_1 x + a_2 b_1 y + a_2 c_1 = 0 \tag{3}$$

$$a_1 a_2 x + a_1 b_2 y + a_1 c_2 = 0. \tag{4}$$

Now *subtract Equation (3) from Equation (4) to produce the equation*

$$(a_1 b_2 - a_2 b_1)y + a_1 c_2 - a_2 c_1 = 0. \tag{5}$$

Now if $a_1 b_2 - a_2 b_1$ is not zero this is a linear equation in y with solution

$$y = \frac{a_2 c_1 - a_1 c_2}{a_1 b_2 - a_2 b_1}. \tag{6}$$

To eliminate y and obtain a linear equation in x alone *we multiply Equation (1) by b_2 and equation (2) by b_1 to produce the equivalent equations*

$$b_2 a_1 x + b_2 b_1 y + b_2 c_1 = 0 \tag{7}$$

$$b_1 a_2 x + b_1 b_2 y + b_1 c_2 = 0. \tag{8}$$

Now subtract Equation (8) from Equation (7)

to produce the equation

$$(b_2 a_1 - b_1 a_2)x + b_2 c_1 - b_1 c_2 = 0. \tag{9}$$

Now if $b_2 a_1 - b_1 a_2$ is not zero this is a linear equation in x with solution

$$x = \frac{b_1 c_2 - b_2 c_1}{b_2 a_1 - b_1 a_2}. \tag{10}$$

Thus (1) and (2) can be solved by the method of elimination and have a unique solution provided

$$a_1 b_2 - a_2 b_1 \neq 0, \tag{11}$$

which is an algebraic representation of the geometrical condition that the lines represented by the equations are neither parallel nor coincident.

If we have a pair of equations in which *one* of the constants a_1, b_1, a_2, b_2 is zero, then we can use the method of solution by substitution. For example, suppose $a_1 = 0$ and b_1, a_2, b_2 are not zero then we have to solve the pair of equations,

$$b_1 y + c_1 = 0 \tag{12}$$

$$a_2 x + b_2 y + c_2 = 0. \tag{2}$$

The first equation in this pair is linear in y and has solution

$$y = -\frac{c_1}{b_1} \quad \text{(since } b_1 \neq 0) \tag{13}$$

We can now substitute this solution into equation (2) to give the equivalent equation

$$a_2 x + b_2 \left(\frac{-c_1}{b_1} \right) + c_2 = 0;$$

that is,

$$a_2 x + \frac{b_1 c_2 - b_2 c_1}{b_1} = 0. \tag{14}$$

Equation (14) is linear in x and has solution

$$x = \frac{b_2 c_1 - b_1 c_2}{a_2 b_1} \quad \text{(since } a_2 \neq 0) \tag{15}$$

Clearly the method of elimination can be combined with the method of substitution to give an alternative method of solution when a_1, b_1, a_2, b_2 are all non-zero; equation (6) combined with Equation (1) or Equation (2) is of the same form as Equation (2) combined with Equation (13).

The method of elimination (with or without substitution) can be extended to a set of three linear equations in three variables. For larger sets of simultaneous equations in a larger number of variables a less laborious technique is required. This technique will be described later in the course.

Exercise 2

(i) Check that

$$x + y + 1 = 0$$
$$2x - y + 2 = 0$$

have a unique solution by using (11), then

(ii) solve the equations by elimination and check the solution by substitution.

Solution 2

(i) $a_1 = 1, b_1 = 1, a_2 = 2, b_2 = -1$;

so

$$a_1 b_2 - a_2 b_1 = -3.$$

Therefore the equations have a unique solution.

(ii) $x + y + 1 = 0$
$2x - y + 2 = 0.$

Add the equations to eliminate y (this is equivalent to multiplying the first equation by 1 and the second by -1 and subtracting) to give

$3x + 3 = 0$

with solution

$x = -1.$

Multiply the first equation by 2 and the second equation by 1 and subtract the equations to give

$3y = 0$

with solution

$y = 0.$

The solution of the pair of equations is then

$x = -1, y = 0.$

Check: first equation gives

$-1 + 0 + 1 = 0$ as required;

second equation gives

$-2 + 0 + 2 = 0$ as required.

1.4 SUMMARY

In the *core material* we have started to develop the language of mathematics by studying the concepts of a set, a mapping and a function. We went on to lay the foundation for our study of the calculus by developing the concepts of a graph, a sequence and a limit.

In the next unit we will concentrate on developing the idea of a function and we will apply the limit concept to functions themselves. In the third unit we will develop a rigorous definition of the limit of a sequence (or of a function) and we will apply this definition to the development of the basic concepts in integral calculus.

In the *optional material* we have seen how the concepts developed in the core material can be extended and applied either to provide a more rigorous mathematical description of a familiar concept (for example our definition of a graph) or to provide a mathematical description of a physical problem (for example the remote pumping engine problem) or to the development of a mathematical technique (an approximate method of solving equations in one variable). We will find in later units that we can come back and examine some of these problems again as we extend and deepen our understanding of the mathematical concepts involved.

When you have completed this unit you should have an understanding of all the terminology and notation used. In addition you should be able to

  (i) evaluate images of a given function,
 (ii) draw the graph of a function,
(iii) find the limit of a given sequence.

Unit 2   Functions and Limits

Contents

2.0 INTRODUCTION

Before reading this introduction you should read through section **1.3.1** (pp. **57–58**).

In the previous unit we introduced the concept of a *function*; in this unit we shall develop the concept a little further. Functions are basic to the development of the mathematics in the course.

In physical situations a functional relationship between two variables can be determined either from a knowledge of some standard physical law (i.e. from theory) or from measurements of the dependent variable for various values of the independent variable (i.e. from experimentation).

Scientists and technologists are often interested in finding a functional relationship which *models** a particular physical situation. Most physical situations are too complicated to be modelled exactly in mathematical terms; for example, we may not know enough about the physical situation to take all the factors into account, or the situation may be too complex for us to do so. Therefore, the model takes into account only the most important aspects of the problem. We shall often use mathematical models in this course, since modelling is the key to the use of mathematics as an aid to problem-solving in science and technology.

In many situations we do not have enough information to apply known theory, so we perform an experiment to determine the functional dependence of one variable on another, and take some measurements. Let us assume for simplicity that our measurements give a set of values of x (the *domain* of the function) and their images, a set of values of y (the *codomain*). The corresponding pairs (x, y) of measurements belong to a subset of the graph of the function, and they give a partial specification of the functional relation between the set of x's and the set of y's. In order to specify our model, we must investigate the form of the functional relationship between the x's and the y's. It could be a linear function

$$x \longmapsto ax + b \qquad (x \in R),$$

a higher order polynomial†

$$x \longmapsto ax^3 + bx^2 + cx + d \qquad (x \in R),$$

a trigonometric function

$$x \longmapsto \sin x \qquad (x \in R),$$

and so on. (Most of the functions we shall meet in this unit have a subset of the real numbers as domain and codomain.)

The choice of the form for the functional relationship will depend on our knowledge of the physical situation from which the data are derived. If we have no idea what sort of function is appropriate, we may draw a graph which fits the data and then try to determine a suitable form for the functional relationship from this.

Having arrived at some decision about the form of the function, we need some method for fitting the curve to the data. But in this unit we are not concerned with methods of curve-fitting; we are interested only in the underlying concept of a function. In science and technology we are often interested in situations where for each measurement of x there is a *unique* measurement of y; that is, where y is a function of x.

The first section of the set book to be studied in this unit is concerned with the manipulation of functions and, in particular, with the operations of addition, multiplication, composition and inversion. Following the pattern of *Unit 1*, we then consider the concept of *limit*, applied to functions. The limit concept and the related concept of

---

* A mathematical model of a physical situation involves replacing physical objects and operations by mathematical ones in such a way that the mathematics takes in to account as accurately as possible relevant facts of the physical situation, and enables us to make predictions about the physical situation.
† The expression $a_n x^n + \cdots + a_1 x + a_0$, $a_n \neq 0$, where n is a positive integer or zero, is called *a polynomial of degree n* and the corresponding function is a *polynomial function of degree n*. A polynomial of degree 1 is said to be *linear*, a polynomial of degree 2 is said to be *quadratic*, etc.

continuity are introduced in preparation for the calculus, which we shall begin to study in *Unit 3*.

In the optional material we begin by considering the decomposition of functions, and then we examine the exponential function and its inverse, the natural logarithm function. As an application of the exponential function we consider a mathematical model for heat exchangers.

The background material for this unit covers the laws of indices and logarithms. If you are already familiar with these concepts, section 2.3 can be omitted entirely.

2.1 CORE MATERIAL

2.1.1 The Arithmetic of Functions

(a) Purpose

To introduce the operations of addition, subtraction, multiplication and division of functions.

(b) Set Book

Study section **1.3.2** (pp. **59–60**) and complete Exercises 1 and 2 (p. **60**).

(c) Notes

The following graphs illustrate the effect of the operations $+$, $-$, $\times$, $\div$ on some simple functions.

The following two figures show the effect of successive multiplication of the function

$$f: x \longmapsto x \qquad (x \in R)$$

by itself.

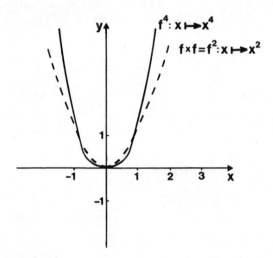

The above figure shows the graphs of $x \longmapsto x^2$ and $x \longmapsto x^4$. Notice that if either of these graphs is reflected in the y-axis it remains unaltered. We say that the y-axis is an *axis of symmetry*. The equation of the reflection of the graph in the y-axis can be expressed by changing x to $-x$ in the original equation. The fact that the graph of a function f remains unaltered can be expressed by the identity

$$f(x) = f(-x)$$

for all x in the domain of f. Such functions are called *even* functions. (The cosine function is another example of an even function.)

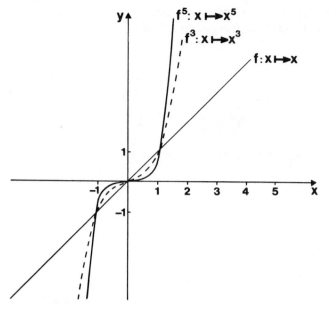

The above figure shows the graphs of

$$x \longmapsto x,$$
$$x \longmapsto x^3,$$
$$x \longmapsto x^5.$$

Here there is a different kind of symmetry. The graph has to be reflected twice, in the y-axis and then in the x-axis, to coincide with itself. Alternatively, we can say we have reflected "through the origin". In general, for any function f we can express this by the identity

$$f(x) = -f(-x)$$

for all x in the domain of f.

In other words, when we reflect such functions in the y-axis (change x to $-x$) we get $-f(x)$ and not $f(x)$. Such functions are called *odd* functions. (The sine function is another example of an odd function.) The identification of symmetry in a problem often helps when deciding what form of function could model a set of data.

(d) Self-Assessment Questions

1   If

$$\left. \begin{array}{l} g: x \longmapsto x^2 + x \\ h: x \longmapsto x^2 - 1 \\ k: x \longmapsto x - 1 \end{array} \right\} \quad (x \in R),$$

express $f: x \longmapsto x^2 + 2x + 1 \quad (x \in R)$ in terms of g, h and k, using $+, -, \times,$ or $\div$.

2   Given

$$\left. \begin{array}{l} f: x \longmapsto x + 1 \\ g: x \longmapsto x \end{array} \right\} \quad (x \in R)$$

find and sketch the graphs of the functions

(i) $f + g \quad (x \in R)$
(ii) $f - g \quad (x \in R)$
(iii) $f \div g \quad (x \in R, x \neq 0)$
(iv) $f \times g \quad (x \in R)$

3   Let

$$\left. \begin{array}{l} g: x \longmapsto a \\ h: x \longmapsto b \end{array} \right\} \quad (x \in R),$$

where a and b are real numbers, $a > 0$ and $b > 1$. If f is any function with domain R, describe how to obtain the graph of the following functions from the graph of f.

(i) $f + g$

(ii) $f - g$

(iii) $h \times f$

(iv) $f \div h$

Solutions

1 $f = g + g - h$. We cannot write $f = g + h/k$ because h/k is not defined at $x = 1$ which is in the domain of f.

2 (i) $f + g : x \longmapsto 2x + 1 \qquad (x \in R)$

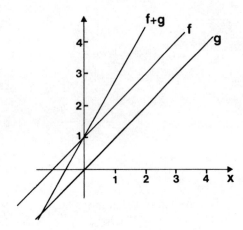

(ii) $f - g : x \longmapsto 1 \qquad (x \in R)$

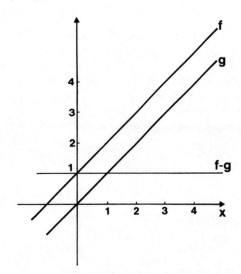

(iii) $\quad f \div g : x \longmapsto 1 + \dfrac{1}{x} \qquad (x \in R, x \neq 0)$

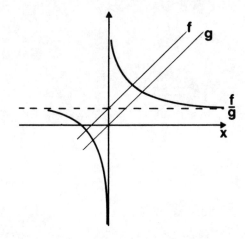

(iv) $\quad f \times g : x \longmapsto x^2 + x \qquad (x \in R)$

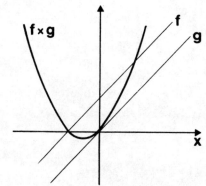

3   (i)   All points on the graph of f are translated in the positive y-direction by the amount a.

   (ii)   As (i), except that points are moved in the negative y-direction by the amount a.

   (iii)   Stretch the vertical scale of each point of the graph of f by the factor b.

   (iv)   Compress the vertical scale of each point of the graph of f by the factor b.

(e)   Terminology

Defined in this section:

sum, difference, product and quotient of functions
odd and even functions.

(f)   Notation

Defined in this section:

$f + g$ The sum of two functions

$$f + g : x \longmapsto f(x) + g(x)$$

$f - g$ The difference of two functions

$$f - g : x \longmapsto f(x) - g(x)$$

$f \times g$ The product of two functions

$$f \times g : x \longmapsto f(x) \times g(x)$$

$f \div g$ The quotient of two functions

$$f \div g : x \longmapsto \frac{f(x)}{g(x)}$$

NOTE: Care must be taken when applying these operations to ensure that f and g have the same domain, and when taking the quotient to ensure that $g(x) \neq 0$ for all x in the domain.

(g) Additional Exercises (OPTIONAL MATERIAL)

1 Given

$$\left. \begin{array}{l} f: x \longmapsto 2^x \\ g: x \longmapsto 2^{x+1} \end{array} \right\} \quad (x \in Z^+)$$

find

(i) $g + f$

(ii) $g \times f$

(iii) $g \div f$

(iv) $g - f$

2 Given

$$\left. \begin{array}{l} s: k \longmapsto 2k - 1 \\ t: k \longmapsto 1/(2k - 1) \\ r: k \longmapsto 2k \end{array} \right\} \quad (x \in Z^+)$$

find

(i) $s \times t$

(ii) $s \div t$

(iii) $s + r$

Solutions

1 (i) $g + f: x \longmapsto 3 \times 2^x$

(ii) $g \times f: x \longmapsto 2^{2x+1}$

(iii) $g \div f: x \longmapsto 2$ $\qquad (x \in Z^+)$

(iv) $g - f: x \longmapsto 2^x$

2 (i) $s \times t: k \longmapsto 1$

(ii) $s \div t: k \longmapsto (2k - 1)^2$ $\qquad (k \in Z^+)$

(iii) $s + r: k \longmapsto 4k - 1$

Notice that, in both these exercises, we could consider the functions as specifying sequences (since they have domain Z^+). Then the combinations of functions suggest natural ways of combining terms of a sequence.

2.1.2 Composition of Functions

(a) Purpose

To introduce *composition* of functions, which is fundamentally different from the arithmetic operations in the previous section.

(b) Set Book

Study section **1.3.3** (pp. **61–64**) and complete Exercises 1 and 2.

(c) Notes

The solution to Exercise 2(i) makes the important point that we can only form the composite function $g \circ f$ when the set of images under f is a subset of the domain of g. We can illustrate this as follows:

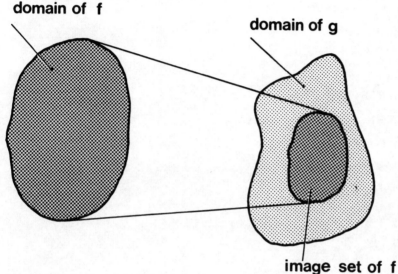

(d) Self-Assessment Questions

1 Given

$$f\colon x \longmapsto x^2 + 2x + 1 \qquad (x \in R)$$

$$g\colon x \longmapsto \frac{1}{x} \qquad (x \in R,\ x \neq 0)$$

then complete the following functions:

$$f \circ g\colon x \longmapsto ?$$
$$g \circ f\colon x \longmapsto ?$$

2 Let

$$f\colon x \longmapsto ax + b \qquad (x \in R)$$

and

$$g\colon x \longmapsto cx + d \qquad (x \in R).$$

Show that $f \circ g$ and $g \circ f$ are linear polynomial functions. Under what conditions does $f \circ g = g \circ f$?

3 Given $f\colon x \longmapsto 2^x \ (x \in R)$, find a function g such that $g \circ f$ is defined and is a linear function.

(Real exponents, like the x in 2^x, are discussed in section 2.3.1.)

Solutions

1 $f \circ g: x \longmapsto \dfrac{1}{x^2} + \dfrac{2}{x} + 1 \qquad (x \in R, x \neq 0).$

$g \circ f$ is not defined because the set of images under f includes 0 (corresponding to $x = -1$) and is therefore not a subset of the domain of g.

2 $f \circ g: x \longmapsto a(cx + d) + b \qquad (x \in R)$

i.e.

$$f \circ g: x \longmapsto acx + ad + b \qquad (x \in R),$$

which is a linear polynomial function.

$$g \circ f: x \longmapsto c(ax + b) + d \qquad (x \in R)$$

i.e.

$$g \circ f: x \longmapsto cax + cb + d \qquad (x \in R),$$

which is again a linear polynomial function.

$g \circ f = f \circ g$ only when $ad + b = cb + d$.

3 Let

$$g: x \longmapsto \log x \qquad (x \in R^+)$$

then

$$g \circ f: x \longmapsto \log 2^x \qquad (x \in R)$$

i.e.

$$g \circ f: x \longmapsto x(\log 2) \qquad (x \in R).$$

This is a linear function.

(It is usual, as here, to write $\log x$ for $\log_{10} x$.)

(e) Terminology

Defined in this section:

composite function

function of a function.

(f) Notation

Defined in this section:

$g \circ f$ The function defined by $g \circ f: x \longmapsto g(f(x))$ ($x \in$ domain of f, and $f(x) \in$ domain of g).

(g) Additional Exercise (OPTIONAL MATERIAL)

1 Additional Exercise 2, section **1.3.6** (p. **76**).

This suggests an alternative way of calculating composite functions.

2 $f: x \longmapsto x^2 + x + 1 \qquad (x \in R)$

$g: x \longmapsto \dfrac{1}{x} \qquad (x \in R, x \neq 0)$

Calculate $f \circ g$ and $g \circ f$, if they exist.

3 Given

$$\left. \begin{array}{l} g: x \longmapsto a \\ h: x \longmapsto b \\ k: x \longmapsto x \end{array} \right\} \qquad (x \in R)$$

where a and b are real numbers; $a > 0$, $b > 1$.

If f is any function with domain R, explain the operation on the graph of f which may be used to obtain the graph of the following:

(i) $f \circ (k + g): x \longmapsto f(x + a)$

(ii) $f \circ (k - g): x \longmapsto f(x - a)$

(iii) $f \circ (h \times k): x \longmapsto f(bx)$

(iv) $f \circ (k \div h): x \longmapsto f\left(\dfrac{x}{b}\right)$

Solutions

2 $f \circ g: x \longmapsto \dfrac{1}{x^2} + \dfrac{1}{x} + 1 \qquad (x \in R, x \neq 0)$

Since $x^2 + x + 1$ is not zero for any $x \in R$, the image set of f is a subset of the domain of g. We can therefore form $g \circ f$

$$g \circ f: x \longmapsto \frac{1}{x^2 + x + 1} \qquad (x \in R)$$

3 (i)   All points on the graph of f are translated in the negative x-direction by the amount a.

(ii) As (i), but points are moved in the positive x-direction by the amount a.

(iii) Compress the scale on the x-axis on the graph of f by the factor b.

(iv) Stretch the scale on the x-axis on the graph of f by the factor b.

Compare this with Self-Assessment Question 3, section 2.1.1(d).

2.1.3   Inverse Functions

(a) Purpose

To investigate reverse mappings and to determine which reverse mappings are inverse functions.

(b) Set Book

Study section **1.3.5** (pp. **66–76**) and work Exercises 1, 2 and 3.

(c) Notes

The point (x, y) on the graph of a mapping f becomes the point (y, x) on the graph of the reverse mapping of f.

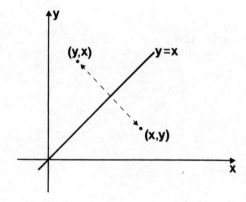

The following examples illustrate that the graph of the reverse mapping of f is the reflection of the graph of f in the line with equation $y = x$. From the first diagram we can see that, although the original mapping is a function, its reverse is not.

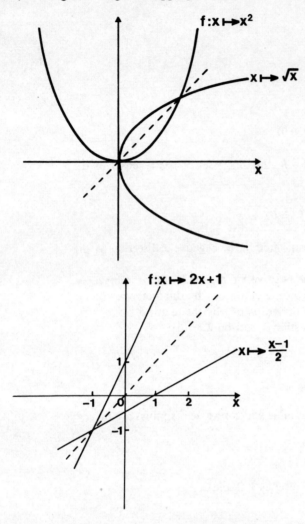

Notice also in the first example, that although the domain of f is R, the domain of the reverse of f is the subset of R for which $x \geqslant 0$.

(d) Self-Assessment Questions

1 Find the reverse mapping for each of the following functions, and say which of these mappings are functions.

(i) $x \longmapsto \dfrac{1}{2}\left(\dfrac{1}{x} - 1\right)$ $(x \in R^+)$

(ii) $x \longmapsto x^2 - 2x + 1$ $(x \in R^+)$

(iii) $x \longmapsto a^x$ $(x \in R)$

where a is a real number > 1.*

---

* In some schoolwork, a^x is taken to be multi-valued; e.g. $4^{1/2} = \pm 2$. If a^x were multi-valued, $x \longmapsto a^x$ would not be a function. See section 2.3.1 where a^x is defined so that $x \longmapsto a^x$ is a function with image set R^+.

2   Sketch the reverse mapping of the function whose graph is given below.

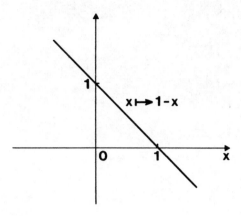

Solutions

1   (i) $x \longmapsto \dfrac{1}{2x + 1}$ $(x \in R, x > -\frac{1}{2})$

The tricky part here is the determination of the domain of the inverse function, that is, the image set of the original function. A quick sketch can be useful. We notice that

(a)   if x is near zero, $\dfrac{1}{x}$ is very large;

(b)   when $x = 1$, the image is zero;

(c)   as x increases beyond 1, $\dfrac{1}{x} - 1$ is negative and $\dfrac{1}{x}$ gets progressively smaller

Thus we have the graph:

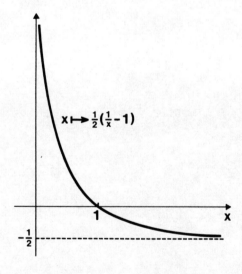

The image set is clearly the set of all real numbers $> -\dfrac{1}{2}$. The reverse mapping is a function.

(ii)   Possibly the simplest way to do this is to notice that

$$x^2 - 2x + 1 = (x - 1)^2$$

15

The reverse mapping is therefore

$$x \longmapsto \{1 + \sqrt{x}, 1 - \sqrt{x}\} \qquad (0 \leqslant x < 1)$$

$$x \longmapsto 1 + \sqrt{x} \qquad\qquad (x \geqslant 1).$$

The reverse mapping is not a function.

(iii)   The graph of the function $x \longmapsto a^x$ is as follows:

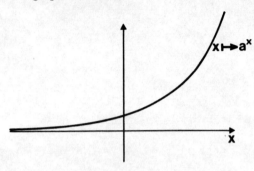

The reverse mapping is $x \longmapsto \log_a x \qquad (x \in R^+)$.

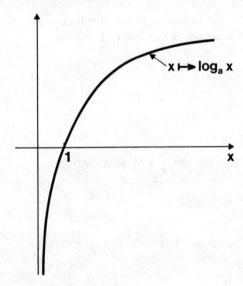

2   The reverse mapping is a function.

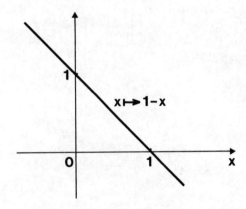

Notice that the function is its own inverse.

Revision

We have given the answers to the first two parts of question 1 in detail: these should be studied carefully if you got any part wrong. See also the answer to Exercise 1 (p. **79**). If you had any difficulty with 2, re-read part (c).

(e) Terminology

Defined in this section:

inverse function
reverse mapping.

(f) Additional Exercises (OPTIONAL MATERIAL)

1 Additional Exercise 1 (p. **76**).

This exercise provides additional practice in the application of some of the terminology used in the earlier parts of this unit.
2 Additional Exercise 4 (p. **77**) gives further practice in the techniques of this section.

2.1.4 Limit of a Function

(a) Purpose

To apply the intuitive concept of a limit to real functions.

(b) Set Book

Study section **1.4.1** (pp. **82–88**) and work Exercises 1 and 2.

(c) Self-Assessment Questions

1 Sketch the graph of each of the following functions and determine the limit of the function (if it exists) for large x in the domain.

(i) $x \longmapsto \sqrt{x}$

(ii) $x \longmapsto 1 + \dfrac{1}{x}$ $\qquad (x \in R^+)$

(iii) $x \longmapsto \dfrac{x}{1 + x}$

2 Determine the limits of the following functions near 2.

(i) $\begin{aligned} x &\longmapsto x^2, \quad x < 2 \\ x &\longmapsto 4, \quad\; x > 2 \end{aligned}$ $\qquad (x \in R^+, x \neq 2)$

(ii) $x \longmapsto \dfrac{x^2 - 4}{x - 2}$ $\qquad (x \in R, x \neq 2)$

3 In our study of calculus we shall often be interested in the limiting behaviour of the quotient $(f(x) - f(a))/(x - a)$ as x approaches a.

(Note that such a quotient is undefined at $x = a$, but this does not prevent the function having a limit near a.) Find

$$\lim_{x \to 1} \frac{f(x) - f(1)}{x - 1}$$

for each of the functions

(i) $f : x \longmapsto x^2$ $\qquad (x \in R)$

(ii) $f : x \longmapsto \dfrac{1}{x}$ $\qquad (x \in R, x \neq 0)$

Solutions

1 (i) No limit.

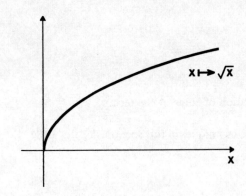

(ii) Limit is 1.

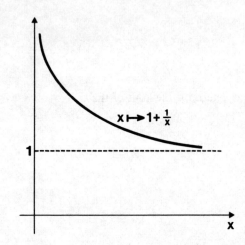

(iii) Limit is 1.

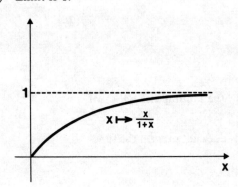

2 (i) Limit is 4.

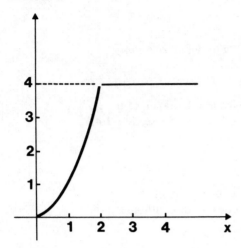

(ii) Limit is 4.

3

(i) $$\frac{f(x)-f(1)}{x-1} = \frac{x^2-1}{x-1}$$

$$= x+1 \qquad (x \in R, x \neq 1),$$

and clearly $x \longmapsto x+1 \qquad (x \in R, x \neq 1)$ has limit 2 near 1.

(ii) $$\frac{f(x)-f(1)}{x-1} = \frac{\dfrac{1}{x}-1}{x-1} = \frac{1-x}{x(x-1)}$$

$$= -\frac{1}{x} \qquad (x \in R, x \neq 0 \text{ and } x \neq 1),$$

and clearly $x \longmapsto -\dfrac{1}{x} \qquad (x \in R, x \neq 0 \text{ and } x \neq 1)$ has limit -1 near 1.

The function

$$x \longmapsto \frac{f(x)-f(a)}{x-a} \qquad (x \in R, x \neq a)$$

is similar to the average velocity function on page **85**. It will play an important role later in the course when we discuss differentiation.

(d) Terminology

Defined in this section:

an intuitive definition of the limit of a function for large x in its domain
an intuitive definition of the limit of a function near a point a.

(e) Notation

Defined in this section:

$\lim_{x \text{ large}} f(x)$ The limit of f for large x in its domain.

$\lim_{x \to a} f(x)$ The limit of f near a.

(f) Additional Exercises (OPTIONAL MATERIAL)

1 Exercise 3 (p. **88**).
2 Additional Exercise 1 (p. **94**).

Both exercises are designed to test and deepen your understanding of the limit concept. The solution to the latter exercise ends with a formal definition of a limit near a point.

2.1.5 Continuity

(a) Purpose

To use the concept of the limit of a function near a point a to define *continuity*.

(b) Set Book

The concept of continuity is very important; many theorems in the branch of mathematics known as *analysis* apply only to continuous functions. Study section **1.4.2** (pp. **89–94**), including Exercise 1.

(c) Self-Assessment Questions

1 List the possible reasons for the discontinuity of a function at a point a.
2 State whether each of the following functions is continuous or discontinuous, and if discontinuous explain why.

$$\text{(i)} \quad x \longmapsto \frac{1}{x^2 - 4} \qquad (x \in R, x \neq \pm 2)$$

$$\text{(ii)} \quad \begin{array}{ll} x \longmapsto x & x < 0 \\ x \longmapsto x^2 & x \geq 0 \end{array} \Bigg\} \quad (x \in R)$$

$$\text{(iii)} \quad \begin{array}{ll} x \longmapsto x^2 & x < 2 \\ x \longmapsto 2 & x \geq 2 \end{array} \Bigg\} \quad (x \in R^+)$$

$$\text{(iv)} \quad \begin{array}{ll} x \longmapsto x^2 + 1 & x \neq 0 \\ x \longmapsto 0 & x = 0 \end{array} \Bigg\} \quad (x \in R)$$

Solutions

1 A function f can be discontinuous at a point a if one of the following conditions holds:

    (i) $f(a)$ is not defined, that is, a is not in the domain of f;
    (ii) $\lim\limits_{x \to a} f(x)$ does not exist;
    (iii) $f(a)$ is defined and $\lim\limits_{x \to a} f(x)$ exists, but $f(a) \neq \lim\limits_{x \to a} f(x)$.

2   (i) The function is discontinuous at ± 2 because it is not defined at these points.
    (ii)

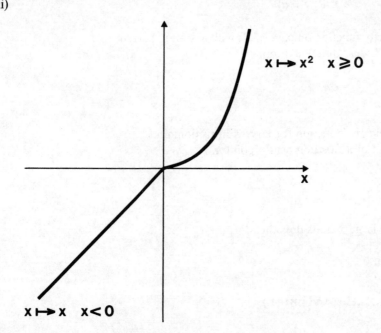

The function is continuous.

(iii)

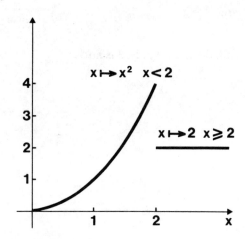

The function is discontinuous at 2 because $\lim_{x \to 2} f(x)$ does not exist.

(iv)

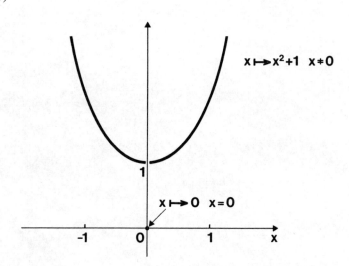

The function is discontinuous at 0 because $\lim_{x \to 0} f(x) \neq f(0)$.

(d)   Terminology

Defined in this section:

continuity
discontinuity
continuous function.

(e)   Additional Exercises (OPTIONAL MATERIAL)

1   Additional Exercise 2 (p. **94**).

This exercise provides an opportunity to revise the ideas of this section.

2   State whether each of the following functions is continuous or discontinuous.

(i) $\begin{aligned} x &\longmapsto x + 1 & x \neq 0 \\ x &\longmapsto 0 & x = 0 \end{aligned}$ $(x \in R)$

(ii) $\begin{aligned} x &\longmapsto 6x^3 & x \geqslant 0 \\ x &\longmapsto 6x & x < 0 \end{aligned}$ $(x \in R)$

(iii) $x \longmapsto \dfrac{x^2 - x - 6}{x - 3}$ $(x \in R, x \neq 3)$

Solutions

2   (i)   The function is discontinuous at 0; $\lim\limits_{x \to 0} f(x) \neq f(0)$.

   (ii)   The function is continuous.

   (iii) The function is discontinuous at 3 because $f(3)$ is not defined; that is, 3 is not in the domain of f.

2.2 OPTIONAL MATERIAL

2.2.1 Decomposition of Functions

(a) Purpose

To explain the process of decomposing a function into simpler functions, and applying this to aid the calculation of images.

(b) Set Book

Study section **1.3.4** (pp. **64–66**) and complete Exercise 1.

2.2.2 The Exponential Function

(a) Purpose

To introduce the exponential function and the natural logarithm function.

(b) Set Book

1 In section **1.5.1** (pp. **100–105**) the exponential function is derived as a model of population growth. This is the subject of the television programme associated with this unit.
2 Study section **1.5.2** (pp. **105–106**) and complete Exercise 1.
  Study section **1.5.3** (pp. **106–107**) and complete Exercise 1.

(c) Notes

1 A function, denoted by exp, is defined as the limit of the sequence

$$1 + x, \left(1 + \frac{x}{2}\right)^2, \left(1 + \frac{x}{3}\right)^3, \ldots \qquad (x \in R)$$

i.e.

$$\exp: x \longmapsto \lim_{k \text{ large}} \left(1 + \frac{x}{k}\right)^k \qquad (k \in Z^+, x \in R).$$

The graph of this function is shown below.

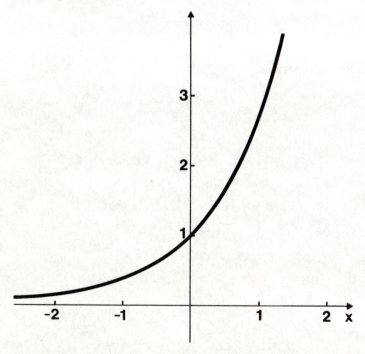

Notice that the domain of exp is R, the image set of exp is R^+ and exp $(0) = 1$; exp $(1) = 2.71828$ to 5 decimal places.

This number, like π, plays a fundamental role in mathematics. Because of its common occurrence in mathematics it is given a special symbol; it is denoted by e.

2 In section **1.5.1** we assume that $\lim\limits_{k \text{ large}} \left(1 + \dfrac{x}{k}\right)^k$ exists for all $x \in R$. In section **1.5.4** this assumption is shown to be true, but you do not need to know how to prove it.

As a corollary of this proof, one of the basic properties of the exponential function is derived, namely

$$\exp(x - y) = \frac{\exp x}{\exp y} \qquad (x, y \in R).$$

From this we find that

(i) if $x - y = z$, then

$$\exp z = \frac{\exp(z + y)}{\exp y}$$

i.e.

$$\exp(z + y) = \exp z \times \exp y$$

(ii) if $x = 0$, then

$$\exp(-y) = \frac{\exp 0}{\exp y}$$

i.e.

$$\exp(-y) = \frac{1}{\exp y}$$

x, y and z conform to the rules of indices (which we revise in section 2.3.1 of this correspondence text).

3 The second proof in section **2.5.4** is the proof of the exponential theorem, i.e.

$$\exp(x) = e^x \qquad (x \in R).$$

Again, the proof of this theorem is not part of the course. It may be useful for you to read through it but you can leave it out if you wish.

The proof demonstrates that

$$\exp x = \left(\exp\left(\frac{x}{k}\right)\right)^k \qquad (x \in R, k \in Z^+).$$

From this equation we can obtain the following properties.

(i) If $x = k = p \qquad (p \in Z^+)$,

$$\exp p = (\exp(1))^p$$
$$= e^p.$$

(ii) If $x = p, k = q \qquad (p, q \in Z^+)$,

$$\exp p = \left(\exp\left(\frac{p}{q}\right)\right)^q.$$

Hence, using (i),

$$e^p = \left(\exp\left(\frac{p}{q}\right)\right)^q.$$

Using the laws of indices this becomes

$$e^{p/q} = \exp p/q,$$

which we can write as

$$e^x = \exp x \qquad (x \in Q^+).$$

This can be extended to negative rationals by using the equation

$$\exp(-x) = \frac{1}{\exp x} \qquad (x \in R) \qquad \text{(from Note (2))}.$$

So if x is a negative rational,

$$\exp(-x) = \frac{1}{e^x} = e^{-x}.$$

Thus far we have established that $\exp x = e^x$ for positive and negative rational values of x. It is also true for $x = 0$ since

$$\exp 0 = 1 \text{ and } e^0 = 1.$$

Therefore

$$\exp x = e^x \qquad (x \in Q).$$

We want to extend this to include all $x \in R$, but what does e^x, x irrational, mean? If x is rational and equals p/q, e^x is the qth root of e^p. The rules of indices cannot help us to define e^x for irrational x. We therefore use the function exp, defined for all $x \in R$, to define e^x for irrational x. We define

$$e^x = \exp x$$

for irrational x, and because this equation already holds for all rational x, it follows that

$$\exp x = e^x \qquad (x \in R).$$

(d) Self-Assessment Questions

1 Sketch the graph of $x \longmapsto e^{-x} \qquad (x \in R)$.
2 Given $y = 2x + 1$, find $\exp y$ in terms of $\exp x$.
3 Given

$$g: \begin{cases} x \longmapsto 1 & x > 0 \\ x \longmapsto 0 & x = 0 \\ x \longmapsto -1 & x < 0 \end{cases} \qquad (x \in R)$$

find the following functions (where possible).

(i) $g \circ \exp$
(ii) $\exp \circ g$
(iii) $\ln \circ \exp$
(iv) $\exp \circ \ln$
(v) $\ln \circ g$
(vi) $g \circ \ln$

Solutions

1

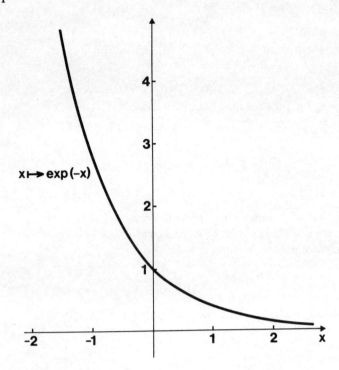

2 $y = 2x + 1$

$\exp y = \exp (2x + 1)$
$\qquad = \exp 2x \exp 1$
$\qquad = (\exp x)^2 e$

3 (i) $g \circ \exp: x \longmapsto 1 \qquad (x \in R)$

(Remember that $\exp x$ is always positive.)

(ii) $\exp \circ g: \begin{cases} x \longmapsto e & x > 0 \\ x \longmapsto 1 & x = 0 \\ x \longmapsto \dfrac{1}{e} & x < 0 \end{cases} \qquad (x \in R)$

(iii) $\ln \circ \exp: x \longmapsto x \qquad (x \in R)$

(iv) $\exp \circ \ln: x \longmapsto x \qquad (x \in R^{+})$

(v) $\ln \circ g$ is not defined, as the set of images under g is not a subset of the domain of ln.

(vi) $g \circ \ln: \begin{cases} x \longmapsto 1 & x > 1 \\ x \longmapsto 0 & x = 1 \\ x \longmapsto -1 & x < 1 \end{cases} \qquad (x \in R^{+})$

Revision

If you had difficulty with question 2, refer back to the basic properties of the exponential function (see Note (1)). If you had difficulty with question 3, then you need to revise the composition of functions (see section **1.3.3**); if you had difficulty with 3(iii) and (iv), then you shold re-read section **1.5.3** and also the solution to Additional Exercise 2 (p. **116**).

(e) Terminology

Defined in this section:

exponential function
natural logarithm function.

(f)   Notation

Defined in this section:

exp   The exponential function defined by

$$\exp: x \longmapsto \lim_{k \text{ large}} \left(1 + \frac{x}{k}\right)^k \qquad (x \in R, k \in Z^+).$$

ln   The natural logarithm function, which is the inverse of the exponential function It has domain R^+.

(g)   Additional Exercise

1   What is the value of

$$\lim_{n \text{ large}} \left(1 + \frac{1}{2n}\right)^n \qquad (n \in Z^+)?$$

Solution

1   $\exp(x) = \displaystyle\lim_{n \text{ large}} \left(1 + \frac{x}{n}\right)^n \qquad (x \in R, n \in Z^+).$

   Hence

$$\exp(\tfrac{1}{2}) = \lim_{n \text{ large}} \left(1 + \frac{1}{2n}\right)^n \qquad (n \in Z^+)$$

$$= e^{1/2} \quad \text{or} \quad \sqrt{e}.$$

2.2.3   Design of a Heat Exchanger

Heat exchangers are components in a wide variety of engineering systems. Familiar examples are the radiator of a motor vehicle (in which hot liquid is cooled by the passage of cold air) and the double cylinder hot tank in a small-bore central heating system (in which cool water is heated by being surrounded by hot water). In this section we shall consider the problem of creating a mathematical model of a heat exchanger in order to apply mathematical techniques to the solution of problems involving the design of heat exchangers.

Let us begin by considering the problem of creating a mathematical model of a constant temperature single-pass tubular heat exchanger, shown diagrammatically in the following figure.

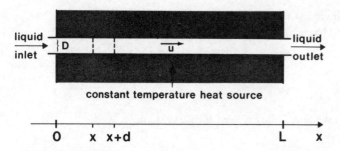

This heat exchanger consists of a cylindrical tube embedded in a block maintained at a constant temperature.

To arrive at our mathematical model, we start with the fundamental principles of heat transfer. In creating our model we shall make use of the following two physical laws (which may be verified experimentally):

> rate at which heat energy is transferred through a surface, *per unit time, per unit surface area* $= H \times$ *(temperature gradient)*,   (1)

where the temperature gradient is the temperature difference across the surface through which heat is being transferred, and where H is a heat transfer constant. The second law is:

heat energy gained = mass of body $\times$ specific heat $\times$ temperature change. (2)

We shall assume that the heat source is at a higher temperature than the liquid in the tube, but if this is not so, a mathematical model can be constructed in a similar manner.

Let us denote the constant temperature of the heat source by S, and the rate of heat energy transfer per unit area into a small section of the fluid by Q. We shall consider the heat transfer into a small section of the fluid, located at point x and of length d as shown in the previous figure. Suppose that the mean temperature inside this section is T; then, applying principle (1), we have

$$Q(\pi Dd) = H(\pi Dd)(S - T),\qquad(3)$$

where D is the diameter of the tube (so that πDd is the surface area of the small section being considered).

In fact, as the second principle tells us, the temperature inside the section is not constant, since the heat energy transferred into the section raises its temperature. If the liquid is flowing at velocity u along the tube, then, applying principle (2), we have

$$Q(\pi Dd) = p\left(\frac{\pi D^2}{4}\right)uC(T_1 - T_0),\qquad(4)$$

where T_0 denotes the temperature of the liquid entering the section, T_1 denotes the temperature of the liquid leaving the section, p denotes the density of the liquid and C denotes the specific heat.

Equating the right-hand sides of Equations (3) and (4) and rearranging we get

$$T_1 = T_0 - \beta(T - S)d,\qquad(5)$$

where $\beta = 4H/(puDC)$.

Equation (5), which gives the change in temperature of a small section, is similar to the population growth equation which we met in section **1.5.1**. We can make Equation (5) exactly similar to the population growth equation by using the approximation $T = T_0,$ and by using the function

$t: x \longmapsto$ *difference in temperature between the heat source and the fluid at x;*

then Equation (5) can be approximated by

$$t(x + d) \simeq (1 - \beta d)t(x).\qquad(6)$$

But, just as in section **1.5.1**, we can sub-divide the interval $[x, x + d]$ and apply Equation (5) together with the approximation $T = T_0$ to each sub-interval. If we make k sub-divisions, we then have

$$t(x + d) \simeq \left(1 - \frac{\beta d}{k}\right)^k t(x).\qquad(7)$$

If we take the limit of the terms in Equation (7) when k is large, then it follows from section **1.5.4** that

$$t(x + d) = e^{-\beta d}t(x).\qquad(8)$$

If we now apply Equation (8) over the whole length of the bar and replace the function t by the function

$T: x \longmapsto$ *temperature of liquid at x,*

we obtain the equation

$$T(x) - S = (T(0) - S)e^{-\beta x}.\qquad(9)$$

If we let

$$\Delta T : x \longmapsto \textit{temperature gradient at the point } x,$$

i.e.

$$\Delta T(x) = T(x) - S,$$

then

$$\Delta T(x) = \Delta T_0 e^{-\beta x}, \tag{10}$$

where $\Delta T_0 = \Delta T(0)$. The graph of the function ΔT is shown below. (In this diagram $\Delta T_1 = \Delta T(L)$.)

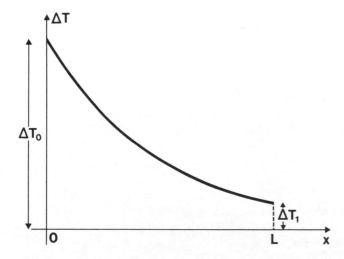

From this result we see that the temperature gradient decreases as the liquid passes through the tube (as we would expect). The rate of decrease depends upon the value of β: the larger the value of β, the faster the rate of decrease. From Equation (5) we see that either decreasing the rate of flow or the diameter of the tube increases β, and thus the rate of decrease of the temperature gradient (as we would expect). Although the results obtained from this model are predictable on an intuitive basis, the same modelling procedure can be applied to other forms of heat exchanger where the results are less predictable.

We can apply the technique described above to two other single pass heat exchangers, in which the heating is performed not by a constant temperature jacket but by a surrounding hot fluid. The arrangement of these forms of heat exchanger is shown in the following figures.

Parallel flow heat exchanger

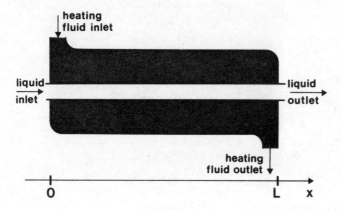

Counter flow heat exhanger

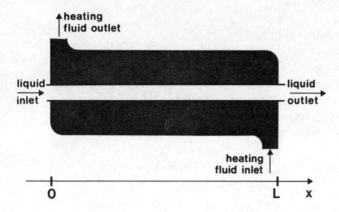

The equations corresponding to Equation (5) for these systems are as follows.

Parallel flow

$$T_1 = T_0 - (\Delta T)(\alpha + \beta)d$$

Counter flow

$$T_1 = T_0 + (\Delta T)(\alpha - \beta)d,$$

where α and β are constants depending on the geometry of the heat exchanger and the nature of the fluids. (β refers to the fluid being heated, as before, and α to the heating fluid.) We can solve these equations, as before, to obtain:

Parallel flow

$$\Delta T(x) = \Delta T_0 \, e^{-(\alpha + \beta)x}$$

Counter flow

$$\Delta T(x) = \Delta T_0 \, e^{-(\beta - \alpha)x}.$$

The parallel flow heat exchanger behaves in a similar manner to the constant temperature heat exchanger, as shown by the following diagram.

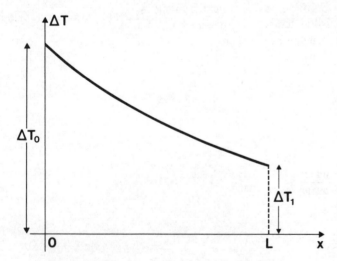

In this case, the rate of decrease of the temperature gradient along the exchanger can be increased by adjusting either α or β or both.

For the counter flow heat exchanger the situation is more complicated. The form of the graph of ΔT depends upon the relative values of α and β, as shown in the following diagrams.

When β > α

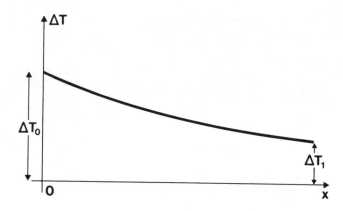

When β = α

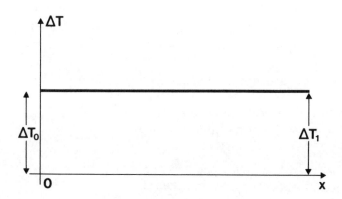

When β < α

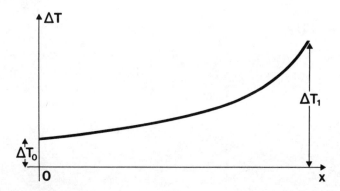

To design an effective heat exchanger we should consider not only the temperature gradient but also the changes in the individual temperatures of the two liquids. It can be shown that these individual temperatures also follow an exponential growth (when heating) or decay (when cooling). To derive these results would involve considerable additional analysis and would produce no new mathematical ideas.

We have seen in this section that the exponential function, which we developed originally to model population growth, can also be used to model temperature change in a simple heat exchanger. In fact, as we shall see later in the course, the exponential function can be used to model a wide variety of growth (and decay) processes.

2.2.4 Functional Iteration

In section 1.2.4 of *Unit 1*, we studied the development of an iterative process for solving equations. Suppose we wish to find a solution of the equation

$$f(x) = 0.$$

This equation can be rewritten as $F(x) = f(x) + x = x$, i.e.

$$F(x) = x.$$

To use an iterative method for solving the equation, we want a recurrence relation, based on $x_k = F(x_{k-1})$, to generate a sequence of successive approximations to the solution of $f(x) = 0$.

We saw in section 1.2.4 that two necessary conditions to ensure that $F(x) = x$ has a solution in the interval $[a, b]$ are

(i) F must have domain and codomain $[a, b]$

and

(ii) F must be continuous in $[a, b]$.

The following problems are still outstanding.

(i) How can we ensure that the interval $[a, b]$ contains only one solution to the equation $f(x) = 0$?

(ii) How can we select a form for $f(x) = 0$ so that the sequence generated by iteration using the corresponding function F is a sequence which converges to the solution of $f(x) = 0$?

We shall now consider this second problem further. The following graph illustrates how (if we can find a suitable rearrangement) a convergent sequence could be obtained.

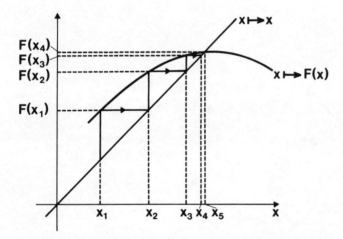

To gain some idea of what determines whether $F(x) = x$ is a suitable rearrangement, we shall simplify the problem by first assuming that the graph of F is a straight line. with domain and codomain R.

Suppose firstly that the graph of F is the straight line $x \longmapsto x + c$, where $c \neq 0$; i.e. a straight line parallel to the graph of $x \longmapsto x$.

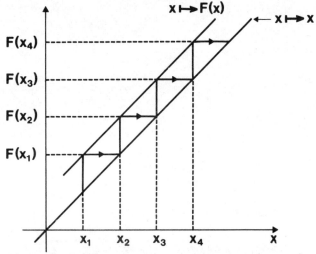

Obviously, the two lines do not intersect, and the sequence $x_1, x_2, x_3, \ldots$ does not converge; i.e. when F has slope 1 the sequence does not converge.

Problem 1

Given that the straight line graph of F has slope s where $0 < s < 1$, show that $F(x) = x$ does produce a sequence converging to the solution of $F(x) = x$.

Solution 1

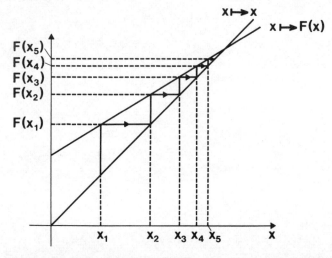

When the slope of F lies in $]0, 1[$* then $F(x) = x$ produces a sequence $x_1, x_2, x_3, \ldots$ converging to the solution of $F(x) = x$.

Problem 2

Given that the slope of the straight line graph of F is zero, show that $F(x) = x$ does produce a convergent sequence.

---

* The reversed brackets indicate an interval without its end-points. Thus
$$]a, b[= \{x \in R : a < x < b\}.$$

Solution 2

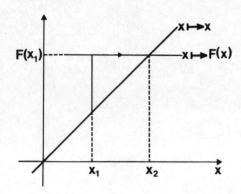

In this case the sequence becomes $x_1, x_2, x_2, x_2, \ldots$ where x_2 is the solution of $F(x) = x$ and so it obviously converges.

We should also consider the cases when the straight line graph of F has slope s where

(i) $s \in {]}{-}1, 0[$
(ii) $s = -1$
(iii) $s < -1$
(iv) $s > 1$

The following diagrams illustrate these four cases:

(i) $s \in {]}{-}1, 0[$

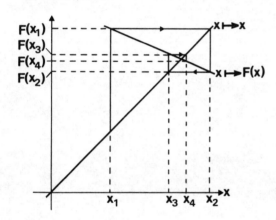

The sequence seems to be converging.

(ii) $s = -1$

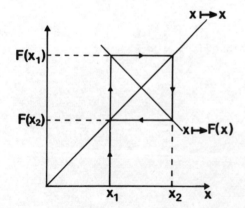

The values of the elements of the sequence alternate between x_1 and x_2, and therefore the sequence does not converge.

(iii) $s < -1$

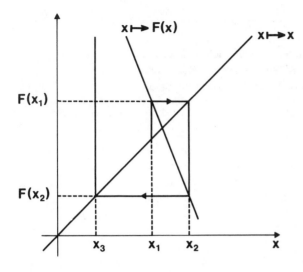

The sequence diverges.

(iv) $s > 1$

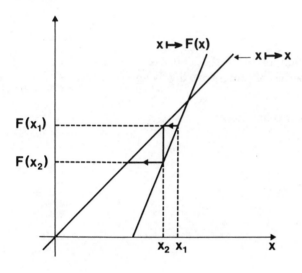

The sequence diverges.

In general, F will not be linear in the inverval $[a, b]$, for if it were we could solve the equation $x = F(x)$ directly. We can, however, use the examples above to suggest a suitable condition to impose on the function F in order to ensure not only that $x = F(x)$ gives rise to a convergent sequence of approximations to a solution of $f(x) = 0$, but also that there is only one solution of $f(x) = 0$ in the interval $[a, b]$. This problem will be considered in later units.

2.3 BACKGROUND MATERIAL

2.3.0 Introduction

The material in this section is not part of the course material. It is included to help you fill gaps in your existing knowledge with relation to your study of the course material. This background material will not feature directly in the assessment and examination.

2.3.1 Laws of Indices

You are familiar with expressions like 3^2, which we read as "3 raised to the power 2" or "3 squared" or "3 to the 2". If a is any real number and n is a positive integer, the product of n a's is written a^n, which we read as "a raised to the power n" or simply "a to the n".

$$\underbrace{a \times a \times a \times \cdots \times a}_{n \text{ factors}} = a^n$$

n is called the *exponent* or power to which a is raised. a^{-n} denotes the reciprocal of a^n for $a \neq 0$.

$$\underbrace{\frac{1}{a \times a \times a \times \cdots \times a}}_{n \text{ factors}} = \frac{1}{a^n} = a^{-n}$$

a^n can be considered as the image of n under the function $x \longmapsto a^x$, $x \neq 0$, $x \in Z$ and $a \neq 0$, $a \in R$. In this section we shall revise the properties of expressions involving a^x for $a \neq 0$, $a \in R$, $x \in Q$.

We start by reminding you of the properties of a^x when x is an integer.

(a) Rule 1

Consider the product $a^m \times a^n$, where m and n are positive or negative integers.

(i) m and n are positive

e.g. $a^2 \times a^3 = (a \times a) \times (a \times a \times a)$

$$= a \times a \times a \times a \times a$$
$$= a^5 = a^{2+3}$$

(ii) m is positive and n is negative, $n \neq -m$

e.g. $a^2 \times a^{-3} = a^2 \times \dfrac{1}{a^3}$

$$= \frac{a \times a}{a \times a \times a} = \frac{1}{a} = a^{-1} = a^{2+(-3)}$$

(iii) m and n are negative

e.g. $a^{-2} \times a^{-3} = \dfrac{1}{a^2} \times \dfrac{1}{a^3} = \dfrac{1}{a \times a} \times \dfrac{1}{a \times a \times a}$

$$= \frac{1}{a^5} = a^{-5} = a^{-2+(-3)}$$

In each case,

$$a^m \times a^n = a^{m+n} \tag{1}$$

The examples only demonstrate the rule. Later in this section the proofs are reproduced for you; read them if you are unconvinced by the examples. What happens when $m = -n$?

$$a^m \times a^n = a^{-n} \times a^n = \frac{a^n}{a^n} = 1$$

But, if (1) holds for all $m, n \in Z$, then

$$a^{-n} \times a^n = a^{-n+n} = a^0.$$

Therefore to ensure that rule (1) holds for all $m, n \in Z$, we must define $a^0 = 1$ for all non-zero real numbers a.

This rule can be extended to the product of any number of terms, e.g.

$$a^m \times a^n \times a^p = a^{m+n+p} \qquad m, n, p \in Z.$$

(b) Rule 2

$(a^m)^n$, where $a \neq 0$, $a \in R$ and $m, n \in Z$, is interpreted as follows. There are again several possible cases to consider.

(i) m and n positive

e.g. $(a^2)^3 = a^2 \times a^2 \times a^2 = a^6 = a^{2 \times 3}$

(ii) m positive and n negative

e.g. $(a^2)^{-3} = \dfrac{1}{(a^2)^3} = \dfrac{1}{a^2 \times a^2 \times a^2} = \dfrac{1}{a^6}$

(iii) m negative and n positive

e.g. $(a^{-2})^3 = a^{-2} \times a^{-2} \times a^{-2} = a^{-6} = a^{-2 \times 3}$

(iv) m and n negative

e.g. $(a^{-2})^{-3} = \dfrac{1}{(a^{-2})^3} = \dfrac{1}{a^{-2} \times a^{-2} \times a^{-2}}$

$$= \dfrac{1}{a^{-6}} = a^6 = a^{-2 \times -3}$$

The examples demonstrate the rule

$$(a^m)^n = a^{mn} \tag{2}$$

A formal proof of this rule will be given later in this section.

(c) Rule 3

$a^m \times b^m = (ab)^m$, $a, b \in R$, $m \in Z$, is easily proved. There are two cases to consider:

(i) m is positive

$$a^m \times b^m = \underbrace{(a \times a \times a \times \cdots \times a)}_{m \text{ factors}} \times \underbrace{(b \times b \times b \times \cdots \times b)}_{m \text{ factors}}$$

$$= \underbrace{(a \times b) \times (a \times b) \times (a \times b) \times \cdots \times (a \times b)}_{m \text{ factors}}$$

So

$$a^m \times b^m = (ab)^m$$

(ii) m is negative

Let $M = -m$, so M is positive.

$$a^m \times b^m = a^{-M} \times b^{-M}$$

$$= \dfrac{1}{a^M} \times \dfrac{1}{b^M}$$

$$= \dfrac{1}{a^M \times b^M}$$

$$= \dfrac{1}{(ab)^M}, \text{ by case (i)}$$

$$= (ab)^{-M}$$

37

So

$$a^m \times b^m = (ab)^m$$

Therefore

$$a^m \times b^m = (ab)^m \qquad a, b \in R, m \in Z \tag{3}$$

(d) Rational Exponents

What does a^m, $m \in Q$ mean? Since m is a rational number, we know it can be written as p/q, $p \in Z$, $q \in Z^+$. We shall consider the special case of $m = 1/q$ when a is positive.

As you know, $\sqrt{2}$ and $\sqrt[3]{6}$ are defined by

$$(\sqrt{2})^2 = 2$$

and

$$(\sqrt[3]{6})^3 = 6$$

In general, if a is positive and n is a positive integer, $\sqrt[n]{a}$ means the *positive* real number whose nth power is a.

If a is negative, however, $\sqrt[n]{a}$ means the real number whose nth power is a. But a negative number does not have real nth roots for all n. For example, there is no real number whose square is -1, for all squares of real numbers are positive or zero. On the other hand, $\sqrt[3]{-1}$ is real: $\sqrt[3]{-1} = -1$.

We shall use the notation $a^{1/n}$ for $\sqrt[n]{a}$.

It can be easily shown that this notation is consistent with the definition of the nth root of a and with the rules just demonstrated.

e.g.

$$(a^{1/q})^q = a$$

Having defined a^m for m of the form $\dfrac{1}{n}$, with n a positive integer, we can now define a^m in the case where m is any rational $\dfrac{p}{q}$. Note that we cannot form a product of $\dfrac{p}{q}$ factors, because, in general, $\dfrac{p}{q}$ is not an integer.

We define $a^{p/q}$ by

$$a^{p/q} = (a^{1/q})^p$$

whenever $a^{1/q}$ is real.

Therefore the rules (1), (2) and (3) can be extended to $m, n \in Q$.

Since finite decimals can be represented as rational numbers, expressions such as $a^{2.5}$ and $10^{11.7}$, in which the exponent or power to which the number is raised is decimal, are meaningful and involve no new concept.

(e) Laws of Indices

We can now state three properties of the expression a^x, $x \in Q$, $a \neq 0$, $a \in R$, known as the laws of indices.

$$\left.\begin{array}{l} a^m \times a^n = a^{m+n} \\ (a^m)^n = a^{mn} \\ a^m \times b^m = (ab)^m \end{array}\right\} \quad \begin{array}{l} m, n \in Z \\ a \neq 0, a \in R \end{array}$$

Notice that $a = 0$ is excluded to allow division by a. When it is meaningful to include $a = 0$, the laws above still hold, and this is very easy to show.

(f) Exercises

1 Simplify the following to the forms indicated.

(i) $\dfrac{(x^n y z^m)^{1/p}}{(x^{1/p} y^p z^m)^n} = x^a y^b z^c$ (i.e. determine a, b and c)

(ii) $[(a^n)^{1/n} x^{1/n}]^{-n} \left[\dfrac{x^n}{a}\right]^{-1/n} = a^b x^y$ (i.e. determine b and y)

2 Simplify $(a^m b^n c^{-p})^m \times \left(\dfrac{c^m}{a^{-m}}\right)^p$; i.e. express it in the form $a^x b^y c^z$ where x, y and z are as simple as possible.

Solutions

1 (i) $\dfrac{(x^n y z^m)^{1/p}}{(x^{1/p} y^p z^m)^n} = \dfrac{(x^n)^{1/p} y^{1/p} (z^m)^{1/p}}{(x^{1/p})^n (y^p)^n (z^m)^n}$

$= \dfrac{x^{n/p} y^{1/p} z^{m/p}}{x^{n/p} y^{pn} z^{mn}}$

Collect x terms, y terms and z terms:

$= x^{n/p - n/p} y^{1/p - pn} z^{m/p - mn}$

$= y^{1/p - pn} z^{m/p - mn}$, since $x^0 = 1$.

(ii) $[(a^n)^{1/n} x^{1/n}]^{-n} \left[\dfrac{x^n}{a}\right]^{-1/n} = [a x^{1/n}]^{-n} \left[\dfrac{x^n}{a}\right]^{-1/n}$

$= a^{-n} (x^{1/n})^{-n} \dfrac{(x^n)^{-1/n}}{a^{-1/n}}$

$= a^{-n} x^{-1} \dfrac{x^{-1}}{a^{-1/n}}$

Collect a and x terms:

$= a^{-n+1/n} x^{-2}$

2 $(a^m b^n c^{-p})^m \times \left(\dfrac{c^m}{a^{-m}}\right)^p = (a^m)^m (b^n)^m (c^{-p})^m \times \dfrac{(c^m)^p}{(a^{-m})^p}$

$= a^{m^2} b^{nm} c^{-pm} \dfrac{c^{mp}}{a^{-mp}}$

Collect a terms, b terms, c terms:

$= a^{m^2 + mp} b^{nm} c^{-pm + mp}$

$= a^{m^2 + mp} b^{nm} c^0$

$= a^{m^2 + mp} b^{nm}$, since $c^0 = 1$.

(g) Proof of Rule 1

Case i) m and n Positive

$a^m \times a^n = \underbrace{(a \times a \times a \times \cdots \times a)}_{m \text{ factors}} \times \underbrace{(a \times a \times a \times \cdots \times a)}_{n \text{ factors}}$

But the right-hand side can be written as

$\underbrace{a \times a \times a \times \cdots \times a}_{(m+n) \text{ factors}} = a^{m+n}$

So

$a^m \times a^n = a^{m+n}$

Case ii)   One of *m* and *n* Positive, the Other Negative

We take *m* positive and *n* negative and let $N = -n$; so N is positive, and $a^m \times a^n = a^m \times a^{-N}$

$$= \frac{\overbrace{a \times a \times a \times \cdots \times a}^{m \text{ factors}}}{\underbrace{a \times a \times a \times \cdots \times a}_{N \text{ factors}}}$$

provided *a* is not zero. We can cancel the common *a*'s. If *m* is greater than *N*

$$a^m \times a^n = \underbrace{a \times a \times a \times \cdots \times a}_{(m-N) \text{ factors}}$$

$$= a^{m-N}$$

So

$$a^m \times a^n = a^{m+n}, \text{ as for case i).}$$

If *m* is less than *N*

$$a^m \times a^n = \frac{1}{\underbrace{a \times a \times a \times \cdots \times a}_{(N-m) \text{ factors}}}$$

$$= a^{-(N-m)}$$

$$= a^{m-N}$$

So

$$a^m \times a^n = a^{m+n}, \text{ as for case i).}$$

Case iii) *m* and *n* Both Negative

Let $M = -m$ and $N = -n$, so *M* and *N* are both positive.

$$a^m \times a^n = a^{-M} \times a^{-N}$$

$$= \frac{1}{a^M} \times \frac{1}{a^N}$$

$$= \frac{1}{a^M \times a^N}$$

$$= \frac{1}{a^{M+N}} \text{ by case i)}$$

$$= a^{-(M+N)}$$

$$= a^{(-M)+(-N)}$$

So

$$a^m \times a^n = a^{m+n}$$

So far, we have seen that if *a* is a non-zero real number and *m* and *n* are integers such that $m + n$ is not zero, then

$$a^m \times a^n = a^{m+n}$$

We have already considered the special case $m = -n$.

(h)   Proof of Rule 2

Case 1) *m* and *n* Positive

$$(a^m)^n = (\underbrace{a \times a \times a \times \cdots \times a}_{m \text{ factors}})^n$$

$$\text{m factors} \qquad \text{m factors} \qquad \text{m factors}$$

$$= \overbrace{(a \times a \times a \times \cdots \times a) \times (a \times a \times a \times \cdots \times a) \times \cdots \times (a \times a \times a \times \cdots \times a)}$$

n factors each containing m factors

$$= \underbrace{a \times a \times a \times \cdots \times a}$$

$m \times n$ factors

So

$$(a^m)^n = a^{mn}$$

Case ii) m Positive, n Negative

Let $N = -n$; so N is positive.

$$(a^m)^n = (a^m)^{-N}$$

$$= \frac{1}{(a^m)^N} \ (a, \text{ remember, is non-zero.})$$

$$= \frac{1}{a^{mN}}, \text{ by case i)}$$

$$= a^{-(mN)}$$

$$= a^{m(-N)}$$

So

$$(a^m)^n = a^{mn}, \text{ as for case i).}$$

Case iii) m Negative, n Positive

Let $M = -m$; so M is positive.

$$(a^m)^n = (a^{-M})^n$$

$$= \left(\frac{1}{a^M}\right)^n \ (a, \text{ remember, is non-zero.})$$

$$= \frac{1}{(a^M)^n}$$

$$= \frac{1}{a^{Mn}}, \text{ by case i)}$$

$$= a^{-Mn}$$

So

$$(a^m)^n = a^{mn}, \text{ as for cases i) and ii).}$$

Case iv) m and n Both Negative

Let $M = -m$ and $N = -n$, so M and N are both positive.

$$(a^m)^n = (a^{-M})^{-N}$$

$$= \left(\frac{1}{a^M}\right)^{-N}$$

$$= \frac{1}{\left(\frac{1}{a^M}\right)^N}$$

$$= \frac{1}{\dfrac{1}{a^{MN}}}, \text{ by case i)}$$

$$= a^{MN}$$

$$= a^{(-m)(-n)}$$

So $(a^m)^n = a^{mn}$

2.3.2 Logarithms

In the core material of this unit we studied reverse mappings and inverse functions. For real positive a we have been careful to define a^x in such a way that the function $x \longmapsto a^x$, $x \in R$, $a \in R^+$, is a one-one function. The inverse function for this function is very important: it is a logarithmic function $f : x \longmapsto \log a^x$ where $x \in R^+$, $a \in R^+$, $a > 1$. We will remind you of the basic properties of the logarithm function.

If x is positive, a is positive and $x = a^n$, where n is some real number, then the number n is called the logarithm to base a of x. We write $n = \log_a x$.

Examples

Base (a)	Number (x)	$\log_a x$ (n)
2	$\frac{1}{16} = 2^{-4}$	-4
2	$1 = 2^0$	0
2	$8 = 2^3$	3
2	$16 = 2^4$	4
4	$16 = 4^2$	2
10	$\frac{1}{10,000} = 10^{-4}$	-4
10	$1 = 10^0$	0
10	$100 = 10^2$	2

Note that logarithms can be positive, zero or negative numbers; and that there is no such thing as *the* logarithm of a number: a base must always be specified. However, in context we do often refer to the logarithm, when the base is clearly understood.

(a) Properties of Logarithms

The Laws of Indices have a corresponding logarithmic form:

	Law	**Logarithm Form**

(i) $a^m \times a^n = a^{m+n}$

$$\log_a (a^m \times a^n) = m + n$$
$$= \log_a a^m + \log_a a^n$$

(ii) $\dfrac{a^m}{a^n} = a^{m-n}$

$$\log_a \left(\frac{a^m}{a^n}\right) = m - n$$
$$= \log_a a^m - \log_a a^n$$

(iii) $(a^m)^n = a^{mn}$

$$\log_a [(a^m)^n] = mn$$
$$= (\log_a a^m) \times n$$

(iv) $(a^m)^{1/n} = a^{m/n}$

$$\log_a [(a^m)^{1/n}] = \frac{m}{n}$$
$$= (\log_a a^m) \div n$$

Example

x	$\log_2 x$
1	0
2	1
4	2
8	3
16	4
32	5
64	6
128	7
256	8
512	9
1024	10
2048	11

The table shows some numbers whose logarithms to base 2 are whole numbers. From law (i), to multiply 64 by 32 we add their logarithms (to base 2), and then use the table to find the product as follows:

x	$\log_2 x$
64	$\longrightarrow$ 6
32	$\longrightarrow$ 5
64×32	$\longrightarrow$ 6 + 5
i.e. 2048	$\longleftarrow$ 11

So that $64 \times 32 = 2048$.

Briefly, then,

"to multiply numbers, add their logarithms".

Exercise 1

By referring to the laws above, complete the following to indicate how to use logarithms to base a.

(i) To divide b by c, subtract _____ from _____.

(ii) To raise b to the power n, multiply _____ by _____.

(iii) To evaluate the nth root of b, divide _____ by _____.

Solution 1

(i) $\log_a c, \log_a b$ (in that order)

(ii) $\log_a b, n$ (*or* $n, \log_a b$)

(iii) $\log_a b, n$ (in that order)

(b) Logarithms to base 10

Logarithms *to base 10*, often called common logarithms, are used for calculations. Tables of such logarithms can be found in sets of mathematical tables and sometimes as an appendix to elementary textbooks.

Here are a few logarithms to base 10.

x	$\log_{10} x$	Comment
$\frac{1}{1000}$	-3	
		All numbers between $\frac{1}{1000}$ and $\frac{1}{100}$ have logarithms between -3 and -2.
$\frac{1}{100}$	-2	
		All numbers between $\frac{1}{100}$ and $\frac{1}{10}$ have logarithms between -2 and -1.
$\frac{1}{10}$	-1	
		All numbers between $\frac{1}{10}$ and 1 have logarithms between -1 and 0.
1	0	
		All numbers between 1 and 0 have logarithms between 0 and 1.
10	1	
		All numbers between 10 and 100 have logarithms between 1 and 2.
100	2	
		All numbers between 100 and 1000 have logarithms between 2 and 3.
1000	3	

8 lies between 1 and 10, and hence its logarithm lies between 0 and 1. In fact $\log_{10} 8 = 0.9031$ (correct to four decimal places).

It is usual to write log x for $\log_{10} x$; in future we shall drop the base 10.

Since any positive number x can be written in the form

$$x = 10^m \times X$$

where m is a positive or negative integer or zero and X is less than 10 and greater than or equal to 1 (i.e., $1 \leqslant X < 10$)

$$\log x = m + \log X.$$

In the list below, in order to preserve the m part (the characteristic) and $\log X$ part (the mantissa), the right-hand column has the two parts separate.

$0.008 = 10^{-3} \times 8$	$\log 0.008 = \log 10^{-3} + \log 8 = -3 + 0.9031$
$0.08 \ = 10^{-2} \times 8$	$\log 0.08 \ = \log 10^{-2} + \log 8 = -2 + 0.9031$
$0.8 \ \ = 10^{-1} \times 8$	$\log 0.8 \ \ = \log 10^{-1} + \log 8 = -1 + 0.9031$
$8 \ \ \ \ = 10^0 \ \ \times 8$	$\log 8 \ \ \ \ = \log 10^0 \ \ + \log 8 = \ \ \ 0 + 0.9031$
$80 \ \ \ = 10^1 \ \ \times 8$	$\log 80 \ \ \ = \log 10^1 \ \ + \log 8 = \ \ \ 1 + 0.9031$
$800 \ \ = 10^2 \ \ \times 8$	$\log 800 \ \ = \log 10^2 \ \ + \log 8 = \ \ \ 2 + 0.9031$
$8000 = 10^3 \ \ \times 8$	$\log 8000 = \log 10^3 \ \ + \log 8 = \ \ \ 3 + 0.9031$

We adopt a notation $\bar{3}$ (read as "bar 3") for -3 and write, for example, $\bar{3}.9031$ to mean $-3 + 0.9031$

i.e.

$\bar{3}.9031$ means the same as -2.0969

(Note that -3.9031 means $-3 - 0.9031$.)

There is no objection to writing -2.0969, but as you will see from our description of the tables, it is more convenient to write $\bar{3}.9031$. After a little practice it is not difficult to perform calculations in this form.

The convenience arises from the fact that the m part (the characteristic) is easily determined, and so it is only necessary for tables of logarithms to present the logarithms of numbers between 1 and 10.

Example

Subtract $\bar{2}.7036$ from 1.9049.

$$
\begin{array}{r}
\bar{1}.9049 - \\
\bar{2}.7036 \\
\hline
1.2013 \\
\end{array}
$$

Note that $\bar{1} - \bar{2} = -1 - (-2) = 1$.

Exercise 2

(i) $\bar{1}.0641$ means
 (a) -1.0641 (b) -0.9359 (c) 0.9359

Write the following numbers in the form $10^m \times X$, $1 \leqslant X < 10$.

(ii) $191.0 = $ _____   (iii) $18.41 = $ _____
(iv) $0.296 = $ _____   (v) $0.0042 = $ _____

Complete the following.

(vi)   log 1.84   = 0.2648; so log 18.4   = _____

(vii) log 4.2   = 0.6232; so log 0.0042 = _____

(viii) log 0.03142 = $\bar{2}$.4972; so log 31.42   = _____

Complete the following using bar notation.

(ix) $\bar{2}$.3078 +
       $\bar{6}$.2941
       _____

(x) $\bar{3}$.4189 −
       $\bar{2}$.6013
       _____

Solution 2

(i)(b)   −0.9359

(ii) $10^2 \times 1.91$

(iii) 10×1.841

(iv) $10^{-1} \times 2.96$

(v) $10^{-3} \times 4.2$

(vi)   1.2648

(vii) $\bar{3}$.6232

(viii) 1.4972

(ix) $\bar{8}$.6019

(x) $\bar{2}$.8176

2.3.3   The Binomial Expansion

Expressions like $(a + b)^0$, $(a + b)^1$, $(a + b)^2$ can easily be written in an expanded form, i.e. without the use of brackets, without much labour; but what about $(a + b)^{59}$? What we are really asking is "Does a general formula exist for the expansion of $(a + b)^n$ where a and b are real numbers and n is a positive integer or zero?"

We shall attempt to discover a general formula by looking at some cases *where n is small*.

(a)   Pascal's Triangle

$n = 0$ $(a + b)^0$ $=$ 1

$n = 1$ $(a + b)^1$ $=$ $a + b$

$n = 2$ $(a + b)^2$ $=$ $a^2 + 2ab + b^2$

$n = 3$ $(a + b)^3$ $=$ $a^3 + 3a^2b + 3ab^2 + b^3$

$n = 4$ $(a + b)^4$ $=$ $a^4 + 4a^3b + 6a^2b^2 + 4ab^3 + b^4$

Note that, in these cases, $(a + b)^n$ has $(n + 1)$ terms when expanded and that we have arranged the terms so that the powers of a decrease and the powers of b increase in each case. With this arrangement we call the terms of $(a + b)^n$:

the first, second, third, ..., $(n + 1)$th, terms;

for example, the fourth term of $(a + b)^4$ is $4ab^3$.

What about $(a + b)^5$? We can guess that it will have 6 terms; we need to determine the numbers that precede the terms, the coefficients of

$a^5, a^4b, a^3b^2, a^2b^3, ab^4, b^5$

Exercise 1

The triangular array of numbers below gives the coefficients of the cases $n = 0, 1, 2, 3$ and 4. By observing the arrowed groups of numbers, which are indicative of a general pattern of the array of numbers,

(i)   guess what the coefficients are for $(a + b)^5$, and hence

(ii) expand $(a + b)^5$

$n = 0$ 1

$n = 1$ 1   1

$n = 2$ 1   2   1

$n = 3$ 1   3   3   1

$n = 4$ 1   4   6   4   1

$n = 5$?

Solution 1

(i) 1 5 10 10 5 1

(ii) $(a + b)^5 = a^5 + 5a^4b + 10a^3b^2 + 10a^2b^3 + 5ab^4 + b^5$

$n = 0$	1					
$n = 1$	1	1				
$n = 2$	1	2	1			
$n = 3$	1	3	3	1		
$n = 4$	1	4	6	4	1	
$n = 5$	1	5	10	10	5	1

Having guessed the $n = 5$ row, let us check it by using the expansion of $(a + b)^4$ given on the previous page:

$$(a + b)^5 = (a + b)(a + b)^4 = (a + b)(a^4 + 4a^3b + 6a^2b^2 + 4ab^3 + b^4)$$
$$= a^5 + 5a^4b + 10a^3b^2 + 10a^2b^3 + 5ab^4 + b^5$$

The triangle of numbers above is called *Pascal's Triangle* [Pascal, 1623–1662], but dates from the middle of the 16th century. Any number of rows can be generated by using the addition property indicated by the arrows on page 45. Using Pascal's Triangle we could, *eventually*, expand $(a + b)^{59}$. But what we really would like is a formula for a general entry in the triangle. We shall seek a formula for the coefficient of the $(r + 1)$th term of $(a + b)^n$, i.e. the coefficient of $a^{n-r}b^r$; this formula must be valid for all positive integers n and for $r = 0, 1, 2, 3, 4, \ldots, n$.

(b) Formula for the Coefficients

We denote the coefficient of $a^{n-r}b^r$ in $(a + b)^n$ by $\binom{n}{r}$ which you can read as "n over r". Thus from Pascal's Triangle, $\binom{6}{2} = 15$, $\binom{4}{3} = 4$, and $\binom{59}{3}$ is the coefficient of $a^{56}b^3$ in $(a + b)^{59}$.

Exercise 2

(i) Use Pascal's Triangle to expand $(a + b)^6$.

(ii) Use Pascal's Triangle to evaluate $\binom{4}{2}$.

(iii) By inspecting any row of Pascal's Triangle, describe any symmetry you see; guess what it might be in terms of $\binom{n}{r}$ and $\binom{n}{n - r}$.

(iv) Inspect column 2 of Pascal's Triangle. What do you think is the entry for the row, $n = 59$?

Solution 2

(i) $(a + b)^6 = a^6 + 6a^5b + 15a^4b^2 + 20a^3b^3 + 15a^2b^4 + 6ab^5 + b^6$

(ii) $\binom{4}{2} = 6$

(iii) The row with $n = 5$, for example, is

```
 ┌──────────┐
 1   5   10   10   5   1
         └────┘
 └──────────────────┘
```

It, like all other rows, is symmetrical, in that the first entry is the same as the last, the second is the same as the last-but-one, and so on.

For $n = 5$ we have

$$\binom{5}{0} = \binom{5}{5}$$

$$\binom{5}{1} = \binom{5}{4}$$

In general

$$\binom{n}{r} = \binom{n}{n-r}$$

(iv)   59

Our aim is to express the coefficient of $a^{n-r}b^r$ in $(a+b)^n$ in terms of n and r

$$\binom{n}{0} \quad \binom{n}{1} \quad \binom{n}{2} \quad \binom{n}{3} \quad \binom{n}{4} \quad \binom{n}{5} \quad \binom{n}{6} \quad \binom{n}{7} \cdots$$

$n = 0$	1						
$n = 1$	1	1					
$n = 2$	1	2	1				
$n = 3$	1	3	3	1			
$n = 4$	1	4	6	4	1		
$n = 5$	1	5	10	10	5	1	
$n = 6$	1	6	15	20	15	6	1
$n = 7$							

$\vdots$

By looking at Pascal's Triangle, we can see that

$$\binom{n}{0} = 1, \text{ for all } n,$$

and that $\binom{n}{1} = n$, for all n.

What about $\binom{n}{2}$? This one is not so obvious but you may notice something about the sequential values of $\dfrac{\binom{n}{2}}{\binom{n}{1}}$ for $n = 2, 3, 4, 5$ and 6.

n	2	3	4	5	6
$\binom{n}{2}$	1	3	6	10	15
$\binom{n}{1}$	2	3	4	5	6
$\dfrac{\binom{n}{2}}{\binom{n}{1}}$	$\frac{1}{2}$	$\frac{3}{3} = \frac{2}{2}$	$\frac{6}{4} = \frac{3}{2}$	$\frac{10}{5} = \frac{4}{2}$	$\frac{15}{6} = \frac{5}{2}$

By always writing the denominator as 2, we notice that the numerator is $n - 1$, so that, in these cases

$$\frac{\binom{n}{2}}{\binom{n}{1}} = \frac{n-1}{2}$$

In fact, this equation holds for all positive integers n; i.e.

$$\binom{n}{2} = \frac{n(n-1)}{2}$$

Exercise 3

(i) Evaluate the sequence obtained by putting $n = 3, 4, 5$ and then 6 in $\dfrac{\binom{n}{3}}{\binom{n}{2}}$ and show

that each ratio equals $\dfrac{n-2}{3}$.

n	3	4	5	6
$\dfrac{\binom{n}{3}}{\binom{n}{2}}$				
$\dfrac{n-2}{3}$				

(ii) Using the fact that $\binom{n}{2} = \dfrac{n(n-1)}{2}$, deduce a general form for $\binom{n}{3}$.

Solution 3

(i)

n	3	4	5	6
$\dfrac{\binom{n}{3}}{\binom{n}{2}}$	$\frac{1}{3}$	$\frac{4}{6}$	$\frac{10}{10}$	$\frac{20}{15}$
$\dfrac{n-2}{3}$	$\frac{1}{3}$	$\frac{2}{3}$	$\frac{3}{3}$	$\frac{4}{3}$

(ii) $\binom{n}{3} = \dfrac{n(n-1)(n-2)}{3 \times 2}$

So far we have

$$\binom{n}{0} = 1,$$

$$\binom{n}{1} = n,$$

$$\binom{n}{2} = \frac{n(n-1)}{2}$$

and

$$\binom{n}{3} = \frac{n(n-1)(n-2)}{2 \times 3}$$

Note the forms of the numerators and denominators for $\binom{n}{2}$ and $\binom{n}{3}$; these suggest, as a guess for $\binom{n}{r}$,

$$\binom{n}{r} = \frac{n(n-1)(n-2)\cdots(n-[r-1])}{2 \times 3 \times 4 \times \cdots \times r}, \qquad (1)$$

This can be made a little more compact with the aid of some more notation.

The product of all the positive integers from n down to 1 is denoted by $n!$ ("n factorial" or "factorial n"); i.e.

$$n! = n(n-1)(n-2)\cdots 1$$

Examples

(i) $6! = 6.5.4.3.2.1 = 720$

(ii) $3! = \dfrac{4!}{4}$

$2! = \dfrac{3!}{3}$

$1! = \dfrac{2!}{2}$

The sequence in example (ii) suggests that we define

$$0! = \frac{1!}{1} = 1.$$

Using the factorial notation, we can write (1) as

$$\binom{n}{r} = \frac{n!}{r! \times (n-r)!} \qquad (2)$$

(2) is the general formula for the coefficient of $a^{n-r}b^r$ in the expansion of $(a+b)^n$, where n is a positive integer or zero and $r = 0, 1, 2, 3, \ldots n$.

(c) The Binomial Expansion for Positive Integral Powers and Zero

Since $(a+b)$ is the sum of two numbers, it is known as a binomial expression.

If a and b are real numbers and n is a positive integer or zero,

$$(a+b)^n = \qquad\qquad (r+1)\text{th term}$$

$$\binom{n}{0}a^n b^0 + \binom{n}{1}a^{n-1}b^1 + \binom{n}{2}a^{n-2}b^2 + \cdots + \binom{n}{r}a^{n-r}b^r + \cdots + \binom{n}{n}a^0 b^n$$

where

$$\binom{n}{r} = \frac{n!}{r!(n-r)!} = \frac{n(n-1)(n-2)\cdots(n-[r-1])}{1 \times 2 \times 3 \times 4 \times \cdots \times r}$$

for $r = 0, 1, 2, \ldots, n$.

Example

$$(1+x)^n = \binom{n}{0}x^0 + \binom{n}{1}x^1 + \binom{n}{2}x^2 + \cdots + \binom{n}{r}x^r + \cdots + \binom{n}{n}x^n$$

$$= 1 + nx + \frac{n(n-1)}{2!}x^2 + \cdots + \frac{n!}{r!(n-r)!}x^r + \cdots + x^n$$

Exercise 4

(i) Expand by the binomial expansion $(1 - x)^m$, quoting the $(r + 1)$th term in full, where $r = 0, 1, 2, \ldots m$.

(ii) Use the binomial expansion to evaluate the coefficient of $a^7 b^3$ in $(a + b)^{10}$.

Solution 4

(i) $(1 - x)^m$

$$= \binom{n}{0} 1^n (-x)^0 + \binom{n}{1} 1^{n-1} (-x)^1 + \binom{n}{2} 1^{n-2} (-x)^2 + \cdots$$

$$+ \binom{n}{r} 1^{n-r} (-x)^r + \cdots + \binom{n}{n} 1^0 (-x)^n$$

$$= 1 - nx + \frac{n(n-1)}{2} x^2 + \cdots + (-1)^r \frac{n!}{r!(n-r)!} x^r + \cdots + (-1)^n x^n$$

(ii) $\binom{10}{3} = \dfrac{10!}{3!(10-3)!} = \dfrac{10 \times 9 \times 8}{3 \times 2 \times 1} = 120$

2.4 SUMMARY

After studying this unit you should have become familiar with the function notation used in this course and have attained a reasonable degree of dexterity in applying the arithmetic rules $(+, -, \times, \div)$ to functions, forming composite functions, and finding inverse functions or reverse mappings. We have intuitively developed the idea of the limit of a function and continuity in preparation for a rigorous treatment in the next unit. These are the important core topics for this unit.

The first optional section is on decomposition of functions. This section is particularly relevant to students interested in computing.

The main optional section this week is section 2.2.2 on the exponential function and the natural logarithm function. The exponential function has many important applications in technology. In section 2.2.3 we model simple heat exchangers as an example of such an application. In a later unit we shall return to this example when we shall have a greater facility for developing the ideas. Although the derivation of the exponential function is optional, all students should be able to use the function $x \longmapsto e^x$ and its inverse $x \longmapsto \ln x$ in examples.

Functional iteration was introduced in the previous correspondence text and in this text we used our knowledge of continuity and limits to develop the idea. We shall return to this example after we have studied differentiation.

In addition to the rigorous treatment of the limit concept already mentioned, in the next unit we introduce the definite integral.

ELEMENTARY MATHEMATICS FOR SCIENCE AND TECHNOLOGY

1 Sets, Mappings and Sequences
2 Functions and Limits
3 The Definite Integral
4 Differentiation
5 The Fundamental Theorem of Calculus
6 NO TEXT
7 Optimization Problems
8 Techniques and Applications of Integration
9 Operations and Relations
10 Taylor Approximation
11 Morphisms–Geometric Vectors
12 NO TEXT
13 Vector Spaces–Matrices
14 Differential Equations
15 Linear Equations
16 Complex Numbers
17 Second Order Differential Equations